T0214456

SpringerBriefs in Probability and Mathematical Statistics

SpringerBriefs present concise summaries of cutting-edge research and practical applications across a wide spectrum of fields. Featuring compact volumes of 50 to 125 pages, the series covers a range of content from professional to academic. Briefs are characterized by fast, global electronic dissemination, standard publishing contracts, standardized manuscript preparation and formatting guidelines, and expedited production schedules.

Typical topics might include:

- A timely report of state-of-the art techniques

- A bridge between new research results, as published in journal articles, and a contextual literature review

- A snapshot of a hot or emerging topic

- Lecture of seminar notes making a specialist topic accessible for non-specialist readers

-SpringerBriefs in Probability and Mathematical Statistics showcase topics of current relevance in the field of probability and mathematical statistics

Manuscripts presenting new results in a classical field, new field, or an emerging topic, or bridges between new results and already published works, are encouraged. This series is intended for mathematicians and other scientists with interest in probability and mathematical statistics. All volumes published in this series undergo a thorough refereeing process.

The SBPMS series is published under the auspices of the Bernoulli Society for Mathematical Statistics and Probability.

More information about this series at http://www.springer.com/series/14353

Benjamin Arras · Christian Houdré

On Stein's Method
for Infinitely Divisible Laws
with Finite First Moment

 Springer

Benjamin Arras
Laboratoire Paul Painlevé
University of Lille Nord de France
Villeneuve-d'Ascq, France

Christian Houdré
School of Mathematics
Georgia Institute of Technology
Atlanta, GA, USA

ISSN 2365-4333   ISSN 2365-4341 (electronic)
SpringerBriefs in Probability and Mathematical Statistics
ISBN 978-3-030-15016-7   ISBN 978-3-030-15017-4 (eBook)
https://doi.org/10.1007/978-3-030-15017-4

Library of Congress Control Number: 2019933697

Mathematics Subject Classification (2010): 60E07, 60E10, 60F05, 47D03, 47D07, 11K65

This Springer imprint is published by the registered company Springer Nature Switzerland AG
The registered company address is: Gewerbestrasse 11, 6330 Cham, Switzerland

To Charlotte and Alice
À CMH

Acknowledgements

Both authors would like to thank Lennart Bondesson for his kind email exchanges on the infinite divisibility of the powers of a normal random variable and to thank Nathan Ross for his comments and bibliographical pointers. Both authors also thank Sorbonne Université for its hospitality while part of this research was carried out. Christian Houdré would like to thank the Institute for Mathematical Sciences of the National University of Singapore for its hospitality and support as well as the organizers of the 2015 Workshop on New Directions in Stein's Method for their invitation. His participation in this workshop ultimately led to the present collaborative work. His research was supported in part by the grants #246283 and #524678 from the Simons Foundation. This material is also based, in part, upon work supported by the National Science Foundation under Grant No. 1440140, while Christian Houdré was in residence at the Mathematical Sciences Research Institute in Berkeley, California, during the fall semester of 2017.

Contents

Abstract

We present, in a unified way, a Stein methodology for infinitely divisible laws (without Gaussian component) having finite first moment. Via a covariance representation, we obtain a characterizing non-local Stein operator which boils down to classical operators in many specific examples. Thanks to this characterizing operator, we introduce various extensions of size-bias and zero-bias distributions and prove that these notions are closely linked to infinite divisibility. When combined with standard Fourier techniques, these extensions also allow to obtain explicit rates of convergence for compound Poisson approximation, in particular, toward the α-stable distribution. Finally, in the setting of nondegenerate self-decomposable laws and for instance stable ones, we solve, by semigroup techniques, the Stein equation induced by the characterizing nonlocal operator. This further leads to quantitative bounds in weak limit theorems, for sums of independent random variables, originating in very classical settings.

Keywords Infinite Divisibility · Self-decomposability · Stable Laws · Stein's Method · Stein–Tikhomirov's Method · Weak Limit Theorems · Rates of Convergence · Kolmogorov Distance · Wasserstein Distance · Smooth Wasserstein Distance

Chapter 1
Introduction

Since its inception in the normal setting (see [85]), Stein's method of approximation has enjoyed tremendous successes in both theory and applications. Starting with Chen's [30] initial extension to the Poisson case the method has been developed for various distributions such as compound Poisson, geometric, negative binomial, exponential, or Laplace, to name but a few. (We refer the reader to Chen, Goldstein and Shao [31] or Ross [81] for good introductions to the method, as well as more precise and complete references.) The methodology developed for the distributions just mentioned is often ad hoc, the fundamental equation changing from one law to another and it is therefore not always easy to see their common underlying thread/approach. There is, however, a large class of random variables for which a common methodology is possible. The class we have in mind is the infinitely divisible one, and it is the purpose of these notes to study Stein's method in this context. Our results will, in particular, provide a common framework for all the examples mentioned above in addition to presenting a bountiful of new ones.

The content of the manuscript is as follows: the next chapter presents background material on the basics, of Stein's method, infinite divisibility, and probability metrics, which are used throughout.

In Chap. 3, Theorem 3.1 provides a functional characterization of infinitely divisible laws from which distance estimates follow. Various comparisons with previously known situations are also made. Then, Corollary 3.5 and Proposition 3.8 respectively extend the notions of size-bias and zero-bias to infinitely divisible distributions whose Lévy measure satisfies moments conditions.

Chapter 4 shows, via Fourier methods, how the new characterization of the previous chapter leads to rates of convergence results, in either Kolmogorov or smooth Wasserstein distance, in particular for the compound Poisson approximations of infinitely divisible distributions.

In Chap. 5, the corresponding Stein's equation is put forward, solved, and its properties studied when the target limiting law is self-decomposable, e.g., α-stable, $1 < \alpha < 2$. Solving the equation associated with such target laws relies mainly on semigroup techniques which are reminiscent of the Gaussian approach developed

© The Author(s), under exclusive license to Springer Nature Switzerland AG 2019
B. Arras and C. Houdré, *On Stein's Method for Infinitely Divisible Laws with Finite First Moment*, SpringerBriefs in Probability and Mathematical Statistics, https://doi.org/10.1007/978-3-030-15017-4_1

in [9, 48]. Chapter 5 ends with Remark 5.9 where Stein's method for the Poisson law, viewed as a particular case of discrete self-decomposable law, is described. This leads to a further extension where a Stein's equation for the convolution between a self-decomposable law and a Poisson law is briefly discussed.

In Chap. 6, the Stein methodology developed to this point is used to obtain quantitative approximation results in classical weak limit theorems for sums of independent random variables. This leads to Theorem 6.2, the main result of the chapter, which is afterward complemented with some more explicit corollaries. In particular, in Theorems 6.7 and 6.8, explicit rates of convergence are first obtained for a limit theorem originating in extreme value theory and then for a nonstandard limit theorem connected to number theory.

These notes end with a brief discussion of further extensions of our ideas and results to be developed in future work.

Chapter 2
Preliminaries

2.1 Stein's Method

Nowadays, Stein's method is a powerful tool to quantify limit theorems appearing in probability theory and, since its introduction in a Gaussian setting, it has been extended to many probability distributions beginning with the Poisson one. The method rests upon basic principles, briefly recalled next, which are in some sense universal and which partially explain its successes.

First, let $Z \sim \mathcal{N}(0, 1)$ be a standard normal random variable and let X be a random variable whose distribution will be compared to the one of Z. It is by now standard that the law of Z is completely characterized via the following covariance identity:

$$\mathbb{E}Zf(Z) = \mathbb{E}f'(Z), \qquad (2.1)$$

since/if the above holds true for all absolutely continuous f for which both $\mathbb{E}|Zf(Z)|$ and $\mathbb{E}|f'(Z)|$ are finite. Then, the main idea of Stein's method goes as follows: the distribution of X is close to the one of Z if, for sufficiently many test functions f,

$$\mathbb{E}Xf(X) \approx \mathbb{E}f'(X).$$

Next, to make this idea more concrete, the maneuver is to introduce an equation capturing the information on the standard normal distribution and, to this end, the fundamental equation is given by

$$f'_{h_1}(x) - xf_{h_1}(x) = h_1(x) - \mathbb{E}h_1(Z), \quad x \in \mathbb{R},$$

where h_1 belongs to an appropriate class of test functions linked to probability metrics of interest. Assuming the existence of a solution for this differential equation and, integrating it with respect to the law of X, provides the following representation:

B. Arras and C. Houdré, *On Stein's Method for Infinitely Divisible Laws with Finite First Moment*, SpringerBriefs in Probability and Mathematical Statistics, https://doi.org/10.1007/978-3-030-15017-4_2

$$|\mathbb{E}h_1(X) - \mathbb{E}h_1(Z)| = \left|\mathbb{E}f'_{h_1}(X) - Xf_{h_1}(X)\right|.$$

This last equality is at the core of Stein's method: the standard normal law is completely encoded into a differential equation so that the problem of bounding the quantity $|\mathbb{E}h_1(X) - \mathbb{E}h_1(Z)|$ only depends on the regularity of the solution f_{h_1}, and on the structure of the law of X. Before going further into the description of the method, let us first recall the following result which makes precise the lines of reasoning presented above. This result is taken from [81, Corollary 3.38].

Theorem 2.1 *Let $Z \sim \mathcal{N}(0, 1)$ and let X be a random variable with finite expectation. Then,*

$$\sup_{x \in \mathbb{R}} |\mathbb{P}(X \leq x) - \mathbb{P}(Z \leq x)| \leq \sup_{f \in \mathcal{F}} \left|\mathbb{E} f'(X) - Xf(X)\right|,$$

where $\mathcal{F}$ is the set of absolutely continuous functions on $\mathbb{R}$ such that $\|f\|_\infty := \sup_{x \in \mathbb{R}} |f(x)| \leq \sqrt{\pi/2}$, $\|f'\|_\infty \leq 2$, and such that for all $u, v, w \in \mathbb{R}$,

$$|(w + u)f(w + u) - (w + v)f(w + v)| \leq \left(|w| + \sqrt{2\pi}/4\right)(|u| + |v|).$$

In the Poisson case, let $W \sim \mathcal{P}(\lambda)$ be a Poisson random variable with parameter $\lambda > 0$ and let Y be a random variable whose distribution is to be compared to the one of W. Now, W is characterized via

$$\mathbb{E}Wf(W) = \lambda\mathbb{E}f(W + 1),$$

since/if this equality holds true for all bounded function f defined on $\mathbb{N}$. Then, again, Y is close, in distribution, to W if, for sufficiently many test functions f,

$$\mathbb{E}Yf(Y) \approx \lambda\mathbb{E}f(Y + 1).$$

Next, the fundamental equation is now a difference equation which is given by

$$\lambda f_{h_2}(n + 1) - nf_{h_2}(n) = h_2(n) - \mathbb{E}h_2(W), \quad n \in \mathbb{N},$$

where now h_2 belongs to an appropriate class of test functions linked to probability metrics of interest. This provides the following representation:

$$|\mathbb{E}h_2(W) - \mathbb{E}h_2(Y)| = \left|\mathbb{E}\lambda f_{h_2}(Y + 1) - Yf_{h_2}(Y)\right|,$$

leading, for example (e.g., see [81, Theorem 4.5]), to:

Theorem 2.2 *Let $W \sim \mathcal{P}(\lambda)$, $\lambda > 0$ and let Y be an integer-valued random variable with mean λ. Then,*

$$\sup_{A\subset\mathbb{N}} |\mathbb{P}\,(Y \in A) - \mathbb{P}\,(W \in A)| \le \sup_{f\in\mathcal{F}} |\mathbb{E}\,\lambda f(Y+1) - Yf(Y)|,$$

where $\mathcal{F}$ is the set of functions such that $\|f\|_\infty := \sup_{n\in\mathbb{N}} |f(n)| \le \min\left(1, \lambda^{-1/2}\right)$ and $\|f(\cdot+1) - f(\cdot)\|_\infty \le (1 - e^{-\lambda})/\lambda \le \min\left(1, \lambda^{-1}\right)$.

As mentioned above, once the fundamental equation is solved, the next step is usually to exploit the regularity of the solution and the structure of the random variables X or Y. Indeed, Stein's method has proven to be fruitful when the following structures are involved: sums of independent random variables, sums of dependent random variables with local dependencies, exchangeable pairs, size-bias (or zero-bias) distributions combined, in a subtle manner, with coupling techniques, regular functionals of random variables (e.g., Malliavin regular functionals). For good literature surveys on the subject as well as standard references, together with persuasive applications, the reader is referred to [10, 13, 27, 28, 31, 35, 63, 72, 81, 86].

Let us briefly describe, next, a way to obtain, from Theorem 2.1, relevant bounds from which explicit rates of convergence are achievable. This method is, by now, classical and is called the bounded zero-bias coupling technique (see [31, Sect. 5.1]). The main point of this approach is to use a distributional transformation linked to the normal distribution together with a bounded coupling. More precisely, let X be a random variable with mean zero and unit variance. The zero-bias transform X^* of X is defined, for all differentiable function f for which the following expectations exist, by:

$$\mathbb{E}Xf(X) = \mathbb{E}f'(X^*).$$

Then, assuming that one can find a coupling of X and X^* such that $|X - X^*| \le \delta$ a.s., for some $\delta > 0$, the following result holds true, e.g., see [31, Theorem 5.1] or [81, Theorem 3.39].

Theorem 2.3 *Let $Z \sim \mathcal{N}(0, 1)$ and let X be a mean zero random variable with unit variance. Let X^* be defined on the same probability space as X and be such that $|X - X^*| \le \delta$ a.s., for some $\delta > 0$. Then,*

$$\sup_{x\in\mathbb{R}} |\mathbb{P}\,(X \le x) - \mathbb{P}\,(Z \le x)| \le \left(1 + \frac{1}{\sqrt{2\pi}} + \frac{\sqrt{2\pi}}{4}\right)\delta. \qquad (2.2)$$

Now, if $X = \sum_{k=1}^{n} X_k/\sqrt{n}$, where the X_k, $k \ge 1$, are i.i.d. zero mean variance one random variables, which moreover are bounded by C, for some $C > 0$ independent of n, then it is possible to build a coupling between X and X^* such that $|X - X^*| \le 2C/\sqrt{n}$. Combined with (2.2), the $1/\sqrt{n}$-rate is then met.

Besides the Gaussian and the Poisson laws, Stein's method has been successfully developed for many univariate distributions such as gamma, Laplace, beta, or semicircular, to name but a few. As put forward in the introduction, the main objective of the current notes is to show that a common Stein methodology is possible for a large class of probability distributions namely, the infinitely divisible one. Indeed, those distributions form a natural class for which a generalized central limit theorem holds and it includes a multitude of well-known distributions, in addition to the Gaussian and the Poisson ones. Moreover, while being naturally linked to limiting theorems for sums of independent summands, a systematic treatment of Stein's method in that setting has not yet appeared in the literature. (Although, as we found out after our first research posting on this topic, this potential subject of investigation is already mentioned by Charles Stein in [86, Lecture XV, Summary, page 158].)

2.2 Infinite Divisibility

Infinite divisibility is a classical subject which has been studied in depth, by many authors, and which is, nowadays, a standard part of probability theory. It was introduced to study the arithmetic of independent random variables and to understand the limiting laws of their sums which was accomplished with great success, e.g., see [45, 59, 62, 66, 77, 84, 87]. Besides its connections to limit theorems, infinite divisibility has also proved fruitful in diverse areas of pure and applied mathematics such as probabilistic number theory [18], analytic number theory [64, 69, 70], Lévy processes and their stochastic calculus as well as their applications to mathematical finance and physics [2, 14, 16, 34]. Let us now recall the definition of infinite divisibility and present some of its main consequences.

Definition 2.4 *A random variable X is infinitely divisible (ID) if, for each $n \geq 1$, there exists $X_{1,n}, \ldots, X_{n,n}$ independent and identically distributed such that*

$$X =_d X_{1,n} + \cdots + X_{n,n}, \tag{2.3}$$

where $=_d$ is short for equality in distribution.

As already mentioned, infinitely divisible distributions are intimately linked to weak limit theorems. Indeed, the set of infinitely divisible distributions is closed with respect to the weak convergence of probability measures (see [84, Lemma 7.8] as well as Remark 4.4 (v) below), while the following is a standard limit theorem for row sums of null arrays (see, e.g., [84, Theorem 9.3]).

Theorem 2.5 *Let $(r_n)_{n \geq 1}$ be a sequence of integers, $r_n \geq 1$, such that $\lim_{n \to +\infty} r_n = +\infty$. Let $\{Z_{n,k} : k = 1, \ldots, r_n, n = 1, 2, \ldots\}$ be such that $Z_{n,1}, \ldots, Z_{n,r_n}, n \geq 1$, are independent and such that, for all $\varepsilon > 0$,*

$$\lim_{n \to +\infty} \max_{1 \leq k \leq r_n} \mathbb{P}\left(\left|Z_{n,k}\right| \geq \varepsilon\right) = 0. \tag{2.4}$$

Let $S_n := \sum_{k=1}^{r_n} Z_{n,k}$, $n \geq 1$. If, for some $c_n \in \mathbb{R}$, $n \geq 1$, the distribution of $S_n + c_n$ converges, as $n \to +\infty$, to a distribution μ, then μ is infinitely divisible.

Recall that the condition (2.4) is equivalent to

$$\lim_{n \to +\infty} \max_{1 \leq k \leq r_n} |\varphi_{Z_{n,k}}(t) - 1| = 0, \tag{2.5}$$

uniformly in t on any compact subset of $\mathbb{R}$, where $\varphi_{Z_{n,k}}$ is the characteristic function of $Z_{n,k}$ (see [77, Lemma 3.1]). Next, and clearly, if φ is the characteristic function of X, then Definition 2.4 is equivalent to the existence, for each $n \geq 1$, of a characteristic function φ_n such that

$$\varphi(t) = (\varphi_n(t))^n, \quad t \in \mathbb{R}.$$

Now, thanks to this fundamental property, the characteristic function of an infinitely divisible random variable has a specific representation which is central to the whole theory. This result, known as the Lévy–Khintchine representation, asserts that X is infinitely divisible if and only if its characteristic function φ is given, for all $t \in \mathbb{R}$, by

$$\varphi(t) = \exp\left(itb - \sigma^2 \frac{t^2}{2} + \int_{-\infty}^{+\infty} (e^{itu} - 1 - itu \mathbb{1}_{|u| \leq 1}) \nu(du) \right), \tag{2.6}$$

for some $b \in \mathbb{R}$, $\sigma \geq 0$ and a positive Borel measure ν on $\mathbb{R}$ such that $\nu(\{0\}) = 0$ and $\int_{-\infty}^{+\infty} (1 \wedge u^2) \nu(du) < +\infty$. This is then abbreviated as $X \sim ID(b, \sigma^2, \nu)$. The measure ν is called the Lévy measure of X, and X is said to be without Gaussian component (or to be purely Poissonian or purely non-Gaussian) whenever $\sigma^2 = 0$. (We refer the reader to Sato [84], for a good introduction to infinitely divisible laws and Lévy processes.) The representation (2.6) is the one which will mainly be used throughout these notes with the (unique) generating triplet (b, σ^2, ν). However, other types of representations are also possible and two of them are presented next. First, if ν is such that $\int_{|u| \leq 1} |u| \nu(du) < +\infty$, then (2.6) becomes

$$\varphi(t) = \exp\left(itb_0 - \sigma^2 \frac{t^2}{2} + \int_{-\infty}^{+\infty} (e^{itu} - 1) \nu(du) \right), \tag{2.7}$$

where $b_0 = b - \int_{|u| \leq 1} u\nu(du)$ is called the *drift* of X. This representation is cryptically expressed as $X \sim ID(b_0, \sigma^2, \nu)_0$. Second, if ν is such that $\int_{|u| > 1} |u| \nu(du) < +\infty$, then (2.6) becomes

$$\varphi(t) = \exp\left(itb_1 - \sigma^2 \frac{t^2}{2} + \int_{-\infty}^{+\infty} (e^{itu} - 1 - itu) \nu(du) \right), \tag{2.8}$$

where $b_1 = b + \int_{|u| > 1} u\nu(du)$ is called the *center* of X. In turn, this last representation is now cryptically written as $X \sim ID(b_1, \sigma^2, \nu)_1$. In fact, $b_1 = \mathbb{E}X$ as, for any $p > 0$,

$\mathbb{E}|X|^p < +\infty$ is equivalent to $\int_{|u|>1} |u|^p \nu(du) < +\infty$. Also, for any $r > 0$, $\mathbb{E}e^{r|X|} < +\infty$ is equivalent to $\int_{|u|>1} e^{r|u|} \nu(du) < +\infty$.

Various choices of generating triplets (b, σ^2, ν) provide various classes of infinitely divisible laws. The triplet $(b, 0, 0)$ corresponds to a degenerate random variable, $(b, \sigma^2, 0)$ to a normal one with mean b and variance σ^2, the choice $(\lambda, 0, \lambda\delta_1)$, where $\lambda > 0$ and where δ_1 is the Dirac measure at 1, corresponds to a Poisson random variable with parameter λ. For ν finite, with the choice $b_0 = b - \int_{|u|\leq 1} u\nu(du) = 0$, $\sigma^2 = 0$ and further setting $\nu(du) = \nu(\mathbb{R})\nu_0(du)$, where ν_0 is a Borel probability measure on $\mathbb{R}$, (2.7) becomes

$$\varphi(t) = \exp\left(\nu(\mathbb{R}) \int_{-\infty}^{+\infty} (e^{itu} - 1)\nu_0(du)\right), \qquad (2.9)$$

i.e., X is compound Poisson: $X \sim CP(\nu(\mathbb{R}), \nu_0)$. Next, let $X \sim \mathcal{N}Bin^0(r, p)$, i.e., let X be negative binomial with support the nonnegative integers and let

$$\mathbb{P}(X = k) = \frac{\Gamma(r+k)}{\Gamma(r)k!} p^r (1-p)^k, \quad k = 0, 1, 2, \dots$$

where $r > 0$ and $0 < p < 1$. Then, $X \sim ID(b, 0, \nu)$ with $\nu(du) = r \sum_{k=1}^{\infty} k^{-1} q^k \delta_k(du)$ and $b_0 = b - \int_{|u|\leq 1} u\nu(du) = 0$, i.e., $b = rq$, and so $\mathbb{E}X = rq/p$, where as usual $q = 1 - p$. If instead, $X \sim \mathcal{N}Bin(r, p)$, i.e., if

$$\mathbb{P}(X = k) = \frac{\Gamma(r+k-1)}{\Gamma(r)(k-1)!} p^r (1-p)^{k-1}, \quad k = 1, 2, \dots$$

then $X \sim ID(b, 0, \nu)$ with $b = 1 + rq$ and $\nu(du) = r \sum_{k=1}^{\infty} k^{-1} q^k \delta_k(du)$ and so $\mathbb{E}X = r/p$. If X has a Gamma distribution with parameters $\alpha > 0$ and $\beta > 0$, i.e., if X has density

$$\beta^\alpha \Gamma(\alpha)^{-1} x^{\alpha-1} e^{-\beta x} 1_{(0,+\infty)}(x), \quad x \in \mathbb{R},$$

then $X \sim ID(b, 0, \nu)$ with $\nu(du) = \alpha e^{-\beta u} u^{-1} 1_{(0,+\infty)}(u)du$ and $b_0 = 0$, i.e., $b = \int_0^1 \alpha e^{-\beta u} du = \alpha(1 - e^{-\beta})/\beta$. If X is the standard Laplace distribution with density $e^{-|x|}/2$, $x \in \mathbb{R}$, then $X \sim ID(b, 0, \nu)$ where $\nu(du) = |u|^{-1} e^{-|u|} du$, $u \neq 0$ and $b_0 = 0$, i.e., $b = \int_{|u|\leq 1} u e^{-|u|} |u|^{-1} du = 0$. More generally, if X has a two-sided exponential distribution with parameters $\alpha > 0$ and $\beta > 0$, i.e., if X has density

$$\alpha\beta(\alpha + \beta)^{-1}(e^{-\alpha x} 1_{[0+\infty)}(x) + e^{\beta x} 1_{(-\infty,0)}(x)), \quad x \in \mathbb{R},$$

then, once more, $X \sim ID(b, 0, \nu)$ with Lévy measure

$$\nu(du) = \left(e^{-\alpha u} u^{-1} 1_{(0,+\infty)}(u) - e^{\beta u} u^{-1} 1_{(-\infty,0)}(u)\right) du,$$

and $b_0 = 0$, i.e., $b = \int_{|u| \leq 1} u \nu(du) = \alpha^{-1}(1 - e^{-\alpha}) - \beta^{-1}(1 - e^{-\beta})$. Finally, if X is a stable random variable, then ν is given by $\nu(B) = \int_{S^0} \sigma(d\xi) \int_0^\infty \mathbb{1}_B(r\xi) \frac{dr}{r^{1+\alpha}}$, $0 < \alpha < 2$, where σ (the spherical component of ν) is a finite positive measure on the unit sphere S^0 of $\mathbb{R}$ ($S^0 = \{-1, 1\}$) and where B is a Borel set of $\mathbb{R}$. Therefore, if X is an α-stable random variable, its Lévy measure is given by

$$\nu(du) := \left(c_1 \frac{1}{u^{\alpha+1}} \mathbb{1}_{(0,+\infty)}(u) + c_2 \frac{1}{|u|^{1+\alpha}} \mathbb{1}_{(-\infty,0)}(u) \right) du, \qquad (2.10)$$

where $c_1, c_2 \geq 0$ are such that $c_1 + c_2 > 0$. The symmetric case corresponds to $c_1 = c_2$ and $b = 0$, which we write as $X \sim S\alpha S$. The class of infinitely divisible distributions is vast and also includes, Student's t-distribution, the Pareto distribution, the F-distribution, the Gumbel distribution to name but a few ones (see [87, B] for more examples). Besides these classical examples let us mention that any log-convex density on $(0, +\infty)$ is infinitely divisible and so are many classes of log-concave measures (see [84, Chap. 10]). It is also noteworthy that the cube or more generally any positive power $q \geq 3$ of a (centered) normal random variable is infinitely divisible. Indeed, when q is a nonnegative integer, i.e., $q \in \mathbb{N}$, such that $q \geq 3$, this is a direct consequence of [19, Theorem 7.3.6] while, for $q \in \mathbb{R}_+ \setminus \mathbb{N}$ with $q \geq 3$, this follows from an adaptation of the proof of [19, Theorem 7.3.6] together with the fact that the probability density function of an α-stable distribution on $(0, +\infty)$ with index $\alpha \leq 1/2$ is hyperbolically completely monotone (see [20]). On the negative side, our framework does not encompass the Stein methodology developed in [40] for the binomial law since, as well known, a nondegenerate bounded random variable cannot be infinitely divisible. For a more recent example of an unbounded continuous law for which a Stein's method has been developed but for which our approach does not apply, see Remark 3.4 (i).

An important subclass of infinitely divisible distributions, which will play a substantial role in our study (see Chap. 5), is formed by the self-decomposable ones. Recall that a random variable X is self-decomposable (SD) if, for any $c \in (0, 1)$, there exists a random variable X_c independent of X such that

$$X =_d cX + X_c. \qquad (2.11)$$

Many of the examples of infinitely divisible distributions with density presented, above, are self-decomposable (again, see Chap. 5, for examples and standard properties of self-decomposability). As illustrated next, self-decomposable distributions are also fundamentally linked to limit theorems for sums of independent uniformly asymptotically negligible summands (see [84, Theorem 15.3]).

Theorem 2.6 *Let μ be a probability measure on $\mathbb{R}$. Let $(Z_k)_{k \geq 1}$ be a sequence of independent random variables and let $b_n > 0$, $n \geq 1$, be such that*

$$\lim_{n \to +\infty} \max_{1 \leq k \leq n} \mathbb{P}\left(|b_n Z_k| \geq \varepsilon \right) = 0. \qquad (2.12)$$

Let $c_n \in \mathbb{R}$, $n \geq 1$, be such that the distribution of $b_n \sum_{k=1}^{n} Z_k + c_n$ converges to μ, as n tends to $+\infty$. Then, μ is self-decomposable. (In case the limit μ is nondegenerate, then necessarily $b_n \to 0$ and $b_{n+1}/b_n \to 1$, $n \to +\infty$.) Conversely, if X is self-decomposable, one can always find $(Z_k)_{k \geq 1}$, $(b_n)_{n \geq 1}$ and $(c_n)_{n \geq 1}$, as above and also satisfying (2.12), such that, as $n \to +\infty$, $(b_n \sum_{k=1}^{n} Z_k + c_n)_{n \geq 1}$ converges in law toward X.

A narrower class of ID laws, still SD, is formed by the stable ones. The characteristic function, φ, of a stable random variable has already been described via its Lévy measure, see (2.10), requiring also that $\sigma^2 = 0$. Another characterization asserts that φ is such that, for any $a > 0$, there exist $b > 0$ and $c \in \mathbb{R}$ satisfying, for all $t \in \mathbb{R}$,

$$\varphi(t)^a = \varphi(bt)e^{ict}.$$

A requirement, more akin to (2.11), for the stability of the random variable X, is that for any $a_1 > 0$, $a_2 > 0$, there exist $a > 0$ and $b \in \mathbb{R}$ such that

$$a_1 X + a_2 X' =_d aX + b,$$

where X' is an independent copy of X. As far as weak convergence, to a stable law, is concerned, the following result is central (see [84, Theorem 15.7]).

Theorem 2.7 *Let μ be a probability measure on $\mathbb{R}$. Let $(Z_k)_{k \geq 1}$ be a sequence of independent and identically distributed random variables. Then, μ is a stable probability measure if and only if there exist $b_n > 0$ and $c_n \in \mathbb{R}$, $n \geq 1$, such that, as n tends to $+\infty$, the distribution of $b_n \sum_{k=1}^{n} Z_k + c_n$ converges toward μ.*

2.3 Some Probability Metrics

At this point, let us introduce various probability metrics, quantifying weak convergence theorems, which are of use throughout these notes. First, very classically, the Kolmogorov distance is:

$$d_K(X, Y) := \sup_{x \in \mathbb{R}} |\mathbb{P}(X \leq x) - \mathbb{P}(Y \leq x)| = \sup_{h \in \mathcal{H}} |\mathbb{E}h(X) - \mathbb{E}h(Y)|,$$

where $\mathcal{H} := \{h : \mathbb{R} \to \mathbb{R}, h = \mathbb{1}_{(-\infty,x]}, x \in \mathbb{R}\}$. In a similar vein, the total variation distance, quantifying Poisson-convergence, is given by:

$$d_{TV}(X, Y) := \sup_{A \in \mathcal{B}(\mathbb{R})} |\mathbb{P}(X \in A) - \mathbb{P}(Y \in A)| = \sup_{h \in \mathcal{H}} |\mathbb{E}h(X) - \mathbb{E}h(Y)|,$$

where $\mathcal{B}(\mathbb{R})$ are the Borel subsets of $\mathbb{R}$, and where $\mathcal{H} := \{h : \mathbb{R} \to \mathbb{R}, h = \mathbb{1}_A, A \in \mathcal{B}(\mathbb{R})\}$. Next, recall that the smooth Wasserstein distance, $d_{W_r}, r \geq 0$, is given by

$$dw_r(X, Y) = \sup_{h \in \mathcal{H}_r} |\mathbb{E}h(X) - \mathbb{E}h(Y)|, \tag{2.13}$$

where $\mathcal{H}_r$ is the set of continuous functions which are r-times continuously differentiable and such that $\|h^{(k)}\|_\infty \le 1$, for all $0 \le k \le r$, where $h^{(0)} = h$, and where $h^{(k)}$, $k \ge 1$, is the kth derivative of h, while $\| \cdot \|_\infty$ is the corresponding supremum norm. At first, note that, by an approximation argument, $d_K(X, Y) \le dw_0(X, Y)$. Next, for $r \ge 1$, another approximation argument (see, e.g., Appendix A of [4] or Lemma A.3 of the Appendix) shows that the smooth Wasserstein distance dw_r can also be represented as

$$dw_r(X, Y) = \sup_{h \in C_c^\infty(\mathbb{R}) \cap \mathcal{H}_r} |\mathbb{E}\,h(X) - \mathbb{E}\,h(Y)|, \tag{2.14}$$

where now, $C_c^\infty(\mathbb{R})$ is the space of compactly supported, infinitely differentiable functions on $\mathbb{R}$. Moreover, by Lemma A.4, for $r \ge 2$,

$$dw_{r-1}(X, Y) \le 3\sqrt{2}\sqrt{dw_r(X, Y)},$$

and, also for $r \ge 2$,

$$dw_1(X, Y) \le \left(3\sqrt{2}\right)^{\sum_{k=1}^{r-1} \frac{1}{2^{k-1}}} \left(dw_r(X, Y)\right)^{\frac{1}{2^{r-1}}}. \tag{2.15}$$

Another very natural distance metrizing the weak convergence of probability measures (see [39]) is the Fortet–Mourier distance given, for any two random variables X and Y, by

$$d_{FM}(X, Y) = \sup_{h \in BLip(1)} |\mathbb{E}h(X) - \mathbb{E}h(Y)|,$$

where $BLip(1)$ is the space of bounded Lipschitz functions, endowed with the norm $\| \cdot \|_{BLip} := \max(\| \cdot \|_\infty, \| \cdot \|_{Lip})$, such that $\| \cdot \|_{BLip} \le 1$, where $\| \cdot \|_{Lip}$ is given, for any h Lipschitz, by $\|h\|_{Lip} := \sup_{x \ne y} |h(x) - h(y)|/|x - y|$. Lemma A.5 of the Appendix ensures that

$$dw_1(X, Y) = d_{FM}(X, Y).$$

In [39], the bounded Lipschitz distance, d_{BL}, is defined as

$$d_{BL}(X, Y) := \sup_{h \in \mathcal{H}} |\mathbb{E}h(X) - \mathbb{E}h(Y)|,$$

where $\mathcal{H}$ is the set of bounded Lipschitz functions h on $\mathbb{R}$ such that $\|h\|_\infty + \|h\|_{Lip} \le 1$. Clearly,

$$d_{BL}(X, Y) \le dw_1(X, Y) \le 2d_{BL}(X, Y),$$

so that both metrics are equivalent, and equivalent to convergence in distribution (see [39, Theorem 11.3.3]). Also, Lemma A.6, via a further approximation argument, ensures that when the law of X has a bounded density h_X with supremum norm $\|h_X\|_\infty$,

$$d_K(X, Y) \leq \left(1 + \frac{\|h_X\|_\infty}{2}\right) \sqrt{d_{W_1}(X, Y)}.$$

This last inequality, combined with (2.15), leads to

$$d_K(X, Y) \leq \left(1 + \frac{\|h_X\|_\infty}{2}\right) \left(3\sqrt{2}\right)^{\sum_{k=1}^{r-1} \frac{1}{2^k}} \left(d_{W_r}(X, Y)\right)^{\frac{1}{2^r}},$$

$r \geq 2$. Furthermore, the smooth Wasserstein distances and the classical Wasserstein distances are ordered in the following way:

$$d_{W_r}(X, Y) \leq d_{W_1}(X, Y) \leq W_1(X, Y) \leq W_p(X, Y), \tag{2.16}$$

for all $r \geq 1$ and where, for any $p \geq 1$, and any two random variables X and Y, each having finite absolute pth moment,

$$W_p^p(X, Y); = \inf \mathbb{E}|X - Y|^p, \tag{2.17}$$

where the infimum is taken over the set of probability measures on $\mathbb{R} \times \mathbb{R}$ with marginals, respectively, given by the law of X and the law of Y. Recall finally that convergence in Wasserstein-p distance is equivalent to convergence in law and convergence of the absolute pth moments (see, e.g., [93, Theorem 6.9]).

In the rest of this text, the terminology Lévy measure is used to denote a positive Borel measure on $\mathbb{R}$ which is atomless at the origin and which integrates out the function $f(x) = \min(1, x^2)$. Moreover, following [84], for a real function f, the terminology "increasing" signifies that $f(s) \leq f(t)$ for $s < t$, while "decreasing" that $f(s) \geq f(t)$ for $s < t$. When equality is not allowed, strictly increasing or strictly decreasing are instead used. We also write ln for the natural logarithm. Finally, all our random variables are assumed to live on the same (rich enough) probability space $(\Omega, \mathcal{F}, \mathbb{P})$.

Chapter 3
Characterization and Coupling

We are now ready to present our first result which characterizes ID laws via a functional equality. This functional equality involves Lipschitz functions and we have to agree on what is meant by "Lipschitz". Below, the functions we consider need not be defined on the whole of $\mathbb{R}$ but just on a subset of $\mathbb{R}$ containing R_X, the range of $X \sim ID(b, 0, \nu)$, and $R_X + S_\nu$, where S_ν is the support of ν. For example, if X is a Poisson random variable, a Lipschitz function (with Lipschitz constant 1) is then defined on $\mathbb{N}$ and is such that $|f(n+1) - f(n)| \leq 1$, for all $n \in \mathbb{N}$.

But, as well known, a Lipschitz function f defined on a subset S of $\mathbb{R}$ can be extended, to the whole of $\mathbb{R}$, without increasing its Lipschitz semi-norm $\|f\|_{Lip}$. (This can be done in various ways, e.g., for any $x \in \mathbb{R}$, let $\tilde{f}(x) = \inf_{z \in S}(f(z) + |x - z|)$. Then, for any $y \in \mathbb{R}$, $\tilde{f}(x) \leq \inf_{z \in S}(f(z) + |x - y| + |y - z|) = |x - y| + \tilde{f}(y)$. Another extension is given via $\bar{f}(x) = \sup_{z \in S}(f(z) - |x - z|)$.) Now, below, in the integral representations and as integrands, f and $\tilde{f}$ are indistinguishable. Therefore, and since we do not wish to distinguish between, say, discrete and continuous random variables, in the sequel, Lipschitz will be understood in the classical sense, i.e., $f \in Lip$ with Lipschitz constant $C > 0$, if $|f(x) - f(y)| \leq C|x - y|$, for all $x, y \in \mathbb{R}$, and f could then be viewed as the Lipschitz extension $\tilde{f}$. Throughout the text, the space of real-valued Lipschitz functions defined on some domain D is denoted by $Lip(D)$, while the space of bounded Lipschitz ones is denoted by $BLip(D)$. Endowed with the norm $\|\cdot\|_{BLip}$, $BLip(D)$ is a Banach space. Finally, we denote the closed unit ball of $Lip(D)$ by $Lip(1)$, and, similarly, $BLip(1)$ denotes the closed unit ball of $BLip(D)$.

Theorem 3.1 *Let X be a random variable such that* $\mathbb{E}|X| < +\infty$. *Let* $b \in \mathbb{R}$ *and let* ν *be a positive Borel measure on* $\mathbb{R}$ *such that* $\nu(\{0\}) = 0$, $\int_{-\infty}^{+\infty}(1 \wedge u^2)\nu(du) < +\infty$ *and* $\int_{|u|>1}|u|\nu(du) < +\infty$. *Then,*

$$\mathbb{E}\left(Xf(X) - bf(X) - \int_{-\infty}^{+\infty}(f(X+u) - f(X)\mathbb{1}_{|u|\leq 1})u\nu(du)\right) = 0, \quad (3.1)$$

© The Author(s), under exclusive license to Springer Nature Switzerland AG 2019
B. Arras and C. Houdré, *On Stein's Method for Infinitely Divisible Laws with Finite First Moment*, SpringerBriefs in Probability and Mathematical Statistics,
https://doi.org/10.1007/978-3-030-15017-4_3

for all bounded Lipschitz function f if and only if $X \sim ID(b, 0, \nu)$.

Proof Note at first that, by the assumption on ν and f, the left-hand side of (3.1) is well defined and that, throughout the proof, interchanges of integrals and expectations are perfectly justified. The direct part of the statement is, in fact, a particular case of a covariance representation obtained in [55, Proposition 2]. Indeed, if $X \sim ID(b, 0, \nu)$ and if f and g are two bounded Lipschitz functions, then

$$Cov(f(X), g(X)) = \int_0^1 \mathbb{E} \int_{-\infty}^{+\infty} (f(X_z + u) - f(X_z))(g(Y_z + u) - g(Y_z))\nu(du)dz,$$

(3.2)

where (X_z, Y_z) is a two-dimensional ID vector with characteristic function defined by $\varphi_z(t, s) = (\varphi(t)\varphi(s))^{1-z}\varphi(t + s)^z$, for all $z \in [0, 1]$, all $s, t \in \mathbb{R}$ and where φ is the characteristic function of X. In other words, $(X_z, Y_z) \sim ID(b_z, 0, \nu_z)$ where $b_z = (b, b)$ and $\nu_z = z\nu_1 + (1 - z)\nu_0$, $z \in [0, 1]$, where $\nu_0(dv, dw) = \nu(dv)\delta_0(dw) + \delta_0(dv)\nu(dw)$ is concentrated on the two main axes of $\mathbb{R}^2$ while $\nu_1(dv, dw)$ is the push-forward of ν to the main diagonal of $\mathbb{R}^2$. Since $X_z =_d Y_z =_d$ X where, again, $=_d$ stands for equality in distribution, taking $g(y) = y$ (which is possible by first taking $g_R(y) = y\mathbb{1}_{|y| \leq R} + R\mathbb{1}_{y \geq R} - R\mathbb{1}_{y \leq -R}$ for $R > 0$ and then passing to the limit), (3.2) becomes

$$\mathbb{E}Xf(X) - \mathbb{E}X\mathbb{E}f(X) = \mathbb{E}\int_{-\infty}^{+\infty} (f(X + u) - f(X))u\nu(du).$$

(3.3)

To pass from (3.3) to (3.1), just note that since $\mathbb{E}|X| < +\infty$, differentiating the characteristic function of X, shows that $\mathbb{E}X = b + \int_{|u|>1} u\nu(du)$. To prove the converse part of the equivalence, i.e., that (3.1), when valid for all bounded Lipschitz functions f, implies that $X \sim ID(b, 0, \nu)$, it is enough to apply (3.1) to sines and cosines or equivalently to complex exponential functions and then to identify the corresponding characteristic function. For any $s \in \mathbb{R}$, let $f(x) = e^{isx}$, $x \in \mathbb{R}$, then (3.1) becomes

$$\mathbb{E}Xe^{isX} - b\mathbb{E}e^{isX} = \mathbb{E}e^{isX}\int_{-\infty}^{+\infty} (e^{isu} - \mathbb{1}_{|u| \leq 1})u\nu(du).$$

(3.4)

Setting $\varphi(s) = \mathbb{E}e^{isX}$, (3.4) rewrites as

$$\varphi'(s) = i\varphi(s)\left(b + \int_{-\infty}^{+\infty} (e^{isu} - \mathbb{1}_{|u| \leq 1})u\nu(du)\right).$$

(3.5)

Integrating out the real and imaginary parts of (3.5) leads, for any $t \geq 0$, to:

$$\varphi(t) = \exp\left(itb + i\int_0^t \int_{-\infty}^{+\infty} (e^{isu} - \mathbb{1}_{|u|\leq 1})u\nu(du)ds\right)$$

$$= \exp\left(itb + i\int_{-\infty}^{+\infty} \int_0^t (e^{isu} - \mathbb{1}_{|u|\leq 1})u\,ds\,\nu(du)\right)$$

$$= \exp\left(itb + \int_{-\infty}^{+\infty} (e^{itu} - 1 - itu\,\mathbb{1}_{|u|\leq 1})\nu(du)\right).$$

A similar computation for $t \leq 0$ finishes the proof. □

Remark 3.2 *(i) Both the statement and the proof of Theorem 3.1 carry over to* $X \sim ID(b, \sigma^2, \nu)$. *The corresponding version of* (3.1) *which characterizes X is then*

$$\mathbb{E}\left(Xf(X) - bf(X) - \sigma^2 f'(X) - \int_{-\infty}^{+\infty} (f(X+u) - f(X)\mathbb{1}_{|u|\leq 1})u\nu(du)\right) = 0.$$
(3.6)

In particular, if $\nu = 0$, (3.6) is the well-known characterization of the normal law with mean $b = \mathbb{E}X$ and variance σ^2.

(ii) There are other ways to restate Theorem 3.1 for X such that $\mathbb{E}|X| < +\infty$. For example, if $X \sim ID(b, 0, \nu)$, then

$$Cov(X, f(X)) = \mathbb{E}\int_{-\infty}^{+\infty} (f(X+u) - f(X))u\nu(du). \qquad (3.7)$$

Conversely, if (3.7) is satisfied for all bounded Lipschitz functions f, then $X \sim ID(b, 0, \nu)$, where $b = \mathbb{E}X - \int_{-\infty}^{+\infty} u\,\mathbb{1}_{|u|>1}\nu(du)$. In case $\int_{|u|\leq 1} |u|\nu(du) < +\infty$, a further characterizing representation is

$$\mathbb{E}Xf(X) - \left(b - \int_{|u|\leq 1} u\nu(du)\right)\mathbb{E}f(X) = \mathbb{E}\int_{-\infty}^{+\infty} f(X+u)u\nu(du), \qquad (3.8)$$

or equivalently,

$$\mathbb{E}Xf(X) - b_0\mathbb{E}f(X) = \mathbb{E}\int_{-\infty}^{+\infty} f(X+u)u\nu(du), \qquad (3.9)$$

i.e.,

$$\mathbb{E}Xf(X) - \left(\mathbb{E}X - \int_{-\infty}^{+\infty} u\nu(du)\right)\mathbb{E}f(X) = \mathbb{E}\int_{-\infty}^{+\infty} f(X+u)u\nu(du). \quad (3.10)$$

Let us now specialize (3.1), (3.7) and (3.9) to various cases, some known and some new.

Examples 3.3 *(i) Of course, if $X \sim ID(\lambda, 0, \lambda\delta_1)$, i.e., when X is a Poisson random variable with parameter $\lambda = \mathbb{E}X > 0$, then (3.1) becomes the familiar*

$$\mathbb{E}Xf(X) = \mathbb{E}X\mathbb{E}f(X + 1).   \tag{3.11}$$

More generally, if $\nu(du) = c\delta_1(du)$, then $X \sim ID(b = \mathbb{E}X, 0, c\delta_1)$.

(ii) If $X \sim \mathcal{N}Bin^0(r, p)$, then, as indicated before, $b_0 = 0$, $\nu(du) = r\sum_{k=1}^{+\infty} q^k$ $\delta_k(du)/k$, with $q = 1 - p$, and so (3.9) becomes

$$\mathbb{E}Xf(X) = r\mathbb{E}\sum_{k=1}^{\infty} f(X + k)q^k$$

$$= rq\mathbb{E}f(X + 1) + r\mathbb{E}\sum_{k=2}^{\infty} f(X + k)q^k$$

$$= rq\mathbb{E}f(X + 1) + \sum_{k=2}^{\infty}\sum_{j=0}^{\infty} f(j + k)r\frac{\Gamma(r + j)}{\Gamma(r)j!}p^r q^j q^k$$

$$= rq\mathbb{E}f(X + 1) + \sum_{\ell=2}^{\infty} f(\ell)p^r q^{\ell}\frac{r}{\Gamma(r)}\sum_{k=0}^{\ell-2}\frac{\Gamma(r + k)}{k!}$$

$$= rq\mathbb{E}f(X + 1) + \sum_{\ell=2}^{\infty} f(\ell)p^r q^{\ell}\frac{1}{\Gamma(r)}\frac{\Gamma(r + \ell - 1)}{(\ell - 2)!}$$

$$= rq\mathbb{E}f(X + 1) + q\mathbb{E}Xf(X + 1),   \tag{3.12}$$

since $\Gamma(t + 1) = t\Gamma(t), t > 0$. Hence, (3.12) is exactly the negative binomial char-acterizing identity obtained in [21].

(iii) If $X \sim \mathcal{N}Bin(r, p)$, (3.1) becomes

$$\mathbb{E}Xf(X) = \mathbb{E}f(X) + r\mathbb{E}\sum_{k=1}^{\infty} f(X + k)q^k$$

$$= \mathbb{E}f(X) + \sum_{k=1}^{\infty}\sum_{j=1}^{\infty} f(j + k)r\frac{\Gamma(r + j - 1)}{\Gamma(r)(j - 1)!}p^r q^{j-1} q^k$$

$$= \mathbb{E}f(X) + \sum_{\ell=2}^{\infty} f(\ell)p^r q^{\ell-1}\frac{r}{\Gamma(r)}\sum_{k=1}^{\ell-1}\frac{\Gamma(r + k - 1)}{(k - 1)!}$$

$$= \mathbb{E}f(X) + \sum_{\ell=2}^{\infty} f(\ell)p^r q^{\ell-1}\frac{1}{\Gamma(r)}\frac{\Gamma(r + \ell - 1)}{(\ell - 2)!}$$

$$= \mathbb{E}f(X) + q\mathbb{E}((r + X - 1)f(X + 1)),   \tag{3.13}$$

which, in view of the previous example, is exactly the expected characterizing identity since $X - 1 \sim \mathcal{N}Bin^0(r, p)$.

(iv) If $X \sim CP(\nu(\mathbb{R}), \nu_0)$, then (3.1) (or (3.8)–(3.10)) becomes

$$\mathbb{E}Xf(X) = \mathbb{E} \int_{-\infty}^{+\infty} f(X+u)u\nu(du) = \nu(\mathbb{R})\mathbb{E} \int_{-\infty}^{+\infty} f(X+u)u\nu_0(du), \quad (3.14)$$

and (3.14) is the characterizing identity for the compound Poisson law given in [11].

(v) If X is the standard Laplace distribution with density $e^{-|x|}/2$, $x \in \mathbb{R}$, then $\nu(du) = |u|^{-1}e^{-|u|}du$, $u \neq 0$, $b = 0$, $\int_{-1}^{1} u\nu(du) = 0$, and (3.1) (or (3.8)–(3.10)) becomes

$$\begin{aligned}
\mathbb{E}Xf(X) &= \mathbb{E} \int_{-\infty}^{+\infty} f(X+u)\mathrm{sign}(u)e^{-|u|}du \\
&= 2\mathbb{E}\left(f(X+L)\,\mathrm{sign}(L)\right) \\
&= \mathbb{E} \int_{0}^{+\infty} (f(X+u) - f(X-u))e^{-u}du \\
&= \mathbb{E}(f(X+Y) - f(X-Y)),
\end{aligned} \quad (3.15)$$

where $\mathrm{sign}(u) = u/|u|$, $u \neq 0$, $\mathrm{sign}(0) = 0$, where L is a standard Laplace random variable independent of X, while Y is a standard exponential random variable independent of X. In [79], a Stein-type operator for the Laplace distribution is introduced as

$$\mathbb{E}f(X) - f(0) = \mathbb{E}f''(X),$$

for f twice differentiable on $\mathbb{R}$ and such that $\|f\|_\infty$, $\|f'\|_\infty$, and $\|f''\|_\infty$ are all finite. Let now, g be a differentiable function on $\mathbb{R}$ such that $\|g\|_\infty < +\infty$ and $\|g'\|_\infty < +\infty$. Then, from (3.15) and Fubini theorem,

$$\begin{aligned}
\mathbb{E}Xg(X) &= \mathbb{E} \int_{0}^{+\infty} (g(X+u) - g(X-u))e^{-u}du \\
&= \mathbb{E} \int_{0}^{+\infty} \left(\int_{-u}^{+u} g'(X+t)dt \right) e^{-u}du \\
&= \mathbb{E} \int_{-\infty}^{+\infty} g'(X+t) \left(\int_{u\geq|t|} e^{-u}du \right) dt \\
&= 2\mathbb{E}\,g'(X+Y),
\end{aligned}$$

where Y is a standard Laplace random variable independent of X.

(vi) If X is a Gamma random variable with parameters $\alpha > 0$ and $\beta > 0$, then, see [37, 43, 65, 78], for f "nice",

$$\mathbb{E}((\beta X - \alpha)f(X)) = \mathbb{E}Xf'(X). \quad (3.16)$$

But, X is infinitely divisible with $\nu(du) = \alpha\mathbb{1}_{(0,+\infty)}(u)\exp(-\beta u)/u\,du$, and it follows from (3.1) that

$$\mathbb{E}Xf(X) = \alpha\frac{1-e^{-\beta}}{\beta}\mathbb{E}f(X)$$

$$+ \alpha\mathbb{E}\int_0^\infty (f(X+u) - f(X)\mathbb{1}_{|u|\leq 1})e^{-\beta u}du. \qquad (3.17)$$

Equivalently from (3.8)–(3.10), since $b_0 = 0$, and since $\mathbb{E}X = \alpha/\beta$,

$$\mathbb{E}Xf(X) = \mathbb{E}\int_{-\infty}^{+\infty} f(X+u)u\nu(du)$$

$$= \alpha\mathbb{E}\int_0^{+\infty} f(X+u)e^{-\beta u}du$$

$$= \frac{\alpha}{\beta}\mathbb{E}f(X+Y)$$

$$= \mathbb{E}X\mathbb{E}f(X+Y),$$

where Y is an exponential random variable, with parameter β, independent of X. Thus, Theorem 3.1 implies the existence of an additive size-bias (see, e.g., [31, Sect. 2]) distribution for the gamma distribution. Moreover, it says that the only probability measure which has an additive exponential size-bias distribution is the gamma one.

(vii) Let $0 < \theta < +\infty$ and let X be a generalized Dickman random variable with parameter θ defined through its characteristic function by

$$\varphi(t) = \exp\left(\theta\int_0^1 \frac{e^{itu}-1}{u}du\right),$$

for all $t \in \mathbb{R}$. Thanks to Theorem 3.1, for all bounded Lipschitz function f

$$\mathbb{E}Xf(X) = \theta\mathbb{E}f(X) + \theta\int_0^1 (f(X+u) - f(X))du$$

$$= \theta\mathbb{E}f(X+U),$$

where U is a uniform random variable on $[0, 1]$ independent of X. Note, in particular, that one recovers the characterizing identity for the generalized Dickman distribution of [5, Chap. 4.2].

(viii) To complement this very partial list, let us consider an example where the literature is sparse (for the symmetric case, see [4, 94]), namely, the stable case. At first, let X be a symmetric α-stable random variable with $\alpha \in (1, 2)$, i.e., let $X \sim S\alpha S$. Then, $b = 0$ and (3.1) becomes

$$\mathbb{E}Xf(X) = \mathbb{E}\int_{-\infty}^{+\infty} \left(f(X+u) - f(X)\mathbb{1}_{\{|u|\leq 1\}}\right)u\nu(du)$$

$$= c\Bigg(-\mathbb{E}\int_{-\infty}^{0}\left(f(X+u) - f(X)\mathbb{1}_{\{|u|\leq 1\}}\right)\frac{du}{(-u)^{\alpha}}$$

$$+\mathbb{E}\int_{0}^{+\infty}\left(f(X+u) - f(X)\mathbb{1}_{\{|u|\leq 1\}}\right)\frac{du}{u^{\alpha}}\Bigg)$$

$$= c\,\mathbb{E}\int_{0}^{+\infty}\left(f(X+u) - f(X-u)\right)\frac{du}{u^{\alpha}},$$

*and, therefore, the previous integral is a fractional operator acting on the test function f. Let us develop this point a bit more by adopting the notation of [83, Sect. 5.4].
The Marchaud fractional derivatives, of order β, of (a sufficiently nice function) f
are defined by*

$$\mathbf{D}_{+}^{\beta}(f)(x) := \frac{\beta}{\Gamma(1-\beta)}\int_{0}^{+\infty}\frac{f(x) - f(x-u)}{u^{1+\beta}}du,$$

$$\mathbf{D}_{-}^{\beta}(f)(x) := \frac{\beta}{\Gamma(1-\beta)}\int_{0}^{+\infty}\frac{f(x) - f(x+u)}{u^{1+\beta}}du.$$

*Note that the above operators are well defined for bounded Lipschitz functions as
soon as β ∈ (0, 1). Then, in a more compact form*

$$\mathbb{E}Xf(X) = C_{\alpha}\mathbb{E}(\mathbf{D}_{+}^{\alpha-1}(f)(X) - \mathbf{D}_{-}^{\alpha-1}(f)(X)), \tag{3.18}$$

*where $C_{\alpha} = c\Gamma(2-\alpha)/(\alpha-1)$. Now, for $X \sim S\alpha S$, [1, Proposition 3.2] or [94,
Theorem 4.1] put forward the following characterizing equation:*

$$\mathbb{E}Xf'(X) = \alpha\mathbb{E}\Delta^{\frac{\alpha}{2}}f(X), \tag{3.19}$$

where $\Delta^{\alpha/2}$ is the fractional Laplacian defined via

$$\Delta^{\frac{\alpha}{2}}f(x) := d_{\alpha}\int_{\mathbb{R}}\frac{f(x+u) - f(x)}{|u|^{1+\alpha}}du,$$

*where $d_{\alpha} = \Gamma(1+\alpha)\sin(\pi\alpha)/(2\pi\cos(\alpha\pi/2))$ and the previous integral has to be
understood as the Cauchy principal value, if it exists, namely,*

$$\int_{\mathbb{R}}\frac{f(x+u) - f(x)}{|u|^{1+\alpha}}du = \lim_{\varepsilon\to 0^{+}}\int_{\mathbb{R}\setminus[-\varepsilon,\varepsilon]}\frac{f(x+u) - f(x)}{|u|^{1+\alpha}}du.$$

*In particular, when the function f is twice continuously differentiable on ℝ with
first and second derivatives bounded, the fractional Laplacian admits the following
representation which is more suited in this situation:*

$$\Delta^{\frac{\alpha}{2}} f(x) = d_\alpha \int_{\mathbb{R}} \frac{f(x+y) - f(x) - yf'(x)}{|y|^{1+\alpha}} dy.$$

On the right-hand side of (3.18), *taking* f' *(nice enough) as a test function leads to*

$$\mathbf{D}_+^{\alpha-1}(f')(x) - \mathbf{D}_-^{\alpha-1}(f')(x) := \frac{\alpha - 1}{\Gamma(2-\alpha)} \int_0^{+\infty} \frac{f'(x+u) - f'(x-u)}{u^\alpha} du.$$

Moreover,

$$\begin{aligned}
\Delta^{\frac{\alpha}{2}} f(x) &= d_\alpha \int_{\mathbb{R}} \frac{f(x+y) - f(x) - yf'(x)}{|y|^{1+\alpha}} dy \\
&= d_\alpha \left(\int_{\mathbb{R}_+} \frac{f(x+y) - f(x) - yf'(x)}{|y|^{1+\alpha}} dy + \int_{\mathbb{R}_-} \frac{f(x+y) - f(x) - yf'(x)}{|y|^{1+\alpha}} dy \right) \\
&= d_\alpha \left(\int_{\mathbb{R}_+} \frac{f(x+y) - f(x) - yf'(x)}{|y|^{1+\alpha}} dy + \int_{\mathbb{R}_+} \frac{f(x-y) - f(x) + yf'(x)}{|y|^{1+\alpha}} dy \right) \\
&= d_\alpha \int_{\mathbb{R}_+} \frac{f(x+y) - f(x) + f(x-y) - f(x)}{|y|^{1+\alpha}} dy \\
&= d_\alpha \int_{\mathbb{R}_+} \left(\int_0^y f'(x+t) - f'(x-t) dt \right) \frac{dy}{|y|^{1+\alpha}} \\
&= \frac{d_\alpha}{\alpha} \int_{\mathbb{R}_+} f'(x+t) - f'(x-t) \frac{dt}{t^\alpha},
\end{aligned} \tag{3.20}$$

showing, for $X \sim S\alpha S$, *the equivalence of the two characterizing identities* (3.18) *and* (3.19). *For the general stable case with Lévy measure given, with* $c_1 \neq c_2$, *by* (2.10) *and with* $b = 0$, *then in a straightforward manner,*

$$\mathbb{E}Xf(X) := c_{2,\alpha}\mathbb{E}(\mathbf{D}_+^{\alpha-1}(f)(X)) - c_{1,\alpha}\mathbb{E}(\mathbf{D}_-^{\alpha-1}(f)(X)) + \frac{(c_1 - c_2)}{\alpha - 1}\mathbb{E}f(X), \tag{3.21}$$

where

$$c_{1,\alpha} = c_1 \frac{\Gamma(2-\alpha)}{\alpha - 1}, \qquad c_{2,\alpha} = c_2 \frac{\Gamma(2-\alpha)}{\alpha - 1}. \tag{3.22}$$

(ix) Another class of infinitely divisible distributions which is of particular interest in a Malliavin calculus framework is the class of second-order Wiener chaoses. As it is well known, if X belongs to this class and if $=_d$ *denotes equality in distribution, then*

$$X =_d \sum_{k=1}^{+\infty} \lambda_k (Z_k^2 - 1),$$

where $(Z_k)_{k\geq 1}$ is a sequence of iid standard normal random variables and where the sequence of reals $(\lambda_k)_{k\geq 1}$ is square summable. Equivalently, the characteristic function of X is given by

$$\varphi(t) = \exp\left(\int_{-\infty}^{+\infty} (e^{itu} - 1 - itu)\nu(du)\right),$$

where

$$\frac{\nu(du)}{du} = \left(\sum_{\lambda\in\Lambda_+} \frac{e^{-u/(2\lambda)}}{2u}\right)\mathbb{1}_{(0,+\infty)}(u) + \left(\sum_{\lambda\in\Lambda_-} \frac{e^{-u/(2\lambda)}}{2(-u)}\right)\mathbb{1}_{(-\infty,0)}(u), \quad (3.23)$$

with $\Lambda_+ = \{\lambda_k : \lambda_k > 0\}$ and $\Lambda_- = \{\lambda_k : \lambda_k < 0\}$. Thus, $X \sim ID(b, 0, \nu)$ with $b = -\int_{|u|>1} u\nu(du)$ and ν as in (3.23). The corresponding characterizing identity is therefore

$$\mathbb{E}Xf(X) = \mathbb{E}\int_{-\infty}^{+\infty} (f(X + u) - f(X))u\nu(du), \quad (3.24)$$

since also $\mathbb{E}X = 0$.

Remark 3.4 *(i) In [68], the authors developed a Stein's method for the two-sided Maxwell distribution whose density is given by $f(x) = x^2 \exp\left(-x^2/2\right)/\sqrt{2\pi}$, for all $x \in \mathbb{R}$. Now, since the Hermite functions are the eigenvectors of the Fourier transform, one can easily compute the characteristic function of the two-sided Maxwell distribution, and it is given by $\varphi(t) = (1 - t^2)\exp(-t^2/2)$, for all $t \in \mathbb{R}$. Since φ vanishes at $t = \pm 1$, the two-sided Maxwell distribution is not infinitely divisible. Moreover, the one-sided Maxwell distribution whose density is equal to $2x^2 \exp(-x^2)/\Gamma(3/2)$, for all $x > 0$, is not infinitely divisible either (see [87, Appendix B, Sect. 3, p 521]).*

(ii) The Tracy–Widom distribution (see, e.g., [90]) is another important distribution for which, unfortunately, our approach does not apply since it is not infinitely divisible (see [38]), and for which a Stein's methodology still needs to be developed.

As a first corollary to Theorem 3.1, the following characterizing identities result extends the notion of additive size-bias distribution to infinitely divisible probability measure with finite nonzero mean and such that $\int_{-1}^{+1} |u|\nu(du) < +\infty$.

Corollary 3.5 *Let X be a nondegenerate random variable such that $\mathbb{E}|X| < +\infty$. Let ν be a Lévy measure such that*

$$\int_{-\infty}^{+\infty} |u|\nu(du) < +\infty, \quad (3.25)$$

and let $b_0 = b - \int_{-1}^{1} u\nu(du)$, $b \in \mathbb{R}$. Assume further that

$$m_0^{\pm} = \max(\pm b_0, 0) + \int_{\mathbb{R}} \tilde{\nu}_{\pm}(du) \neq 0 \qquad (3.26)$$

with $\tilde{\nu}(du) = u\nu(du)$. Then,

$$\mathbb{E}Xf(X) = m_0^{+}\mathbb{E}f(X + Y^{+}) - m_0^{-}\mathbb{E}f(X + Y^{-}), \qquad (3.27)$$

for all bounded Lipschitz functions f, where the random variables Y^{+}, Y^{-} and X are independent with Y^{+} and Y^{-} having respective law

$$\mu_{Y^{\pm}}(du) = \frac{b_0^{\pm}}{m_0^{\pm}}\delta_0(du) + \frac{\tilde{\nu}_{\pm}(du)}{m_0^{\pm}}, \qquad (3.28)$$

with $b_0^{+} = \max(b_0, 0)$ and $b_0^{-} = -\min(b_0, 0)$, if and only if $X \sim ID(b, 0, \nu)$.

Proof Let f be a bounded Lipschitz function. By (3.9),

$$\mathbb{E}Xf(X) = b_0\mathbb{E}f(X) + \mathbb{E}\int_{-\infty}^{+\infty} f(X + u)u\nu(du). \qquad (3.29)$$

Now since $\mathbb{E}|X| < +\infty$ and thanks to (3.25), $\tilde{\nu}(du) = u\nu(du)$ is a finite signed measure and so its Jordan decomposition is given by

$$\tilde{\nu}(du) = \tilde{\nu}_{+}(du) - \tilde{\nu}_{-}(du) = u\mathbb{1}_{(0,+\infty)}(u)\nu(du) - (-u)\mathbb{1}_{(-\infty,0)}(u)\nu(du).$$

Therefore, (3.29) becomes

$$\mathbb{E}Xf(X) = b_0^{+}\mathbb{E}f(X) - b_0^{-}\mathbb{E}f(X) + \mathbb{E}\int_0^{+\infty} f(X + u)\tilde{\nu}_{+}(du) - \mathbb{E}\int_{-\infty}^{0} f(X + u)\tilde{\nu}_{-}(du).$$

Now,

$$m_0^{+} = b_0^{+} + \int_{\mathbb{R}} \tilde{\nu}_{+}(du) = b_0^{+} + \int_0^{+\infty} u\nu(du),$$

$$m_0^{-} = b_0^{-} + \int_{\mathbb{R}} \tilde{\nu}_{-}(du) = b_0^{-} + \int_{-\infty}^{0} (-u)\nu(du),$$

($b_0 = b_0^{+} - b_0^{-}$ and $m_0^{+} - m_0^{-} = \mathbb{E}X$) and, therefore, introducing the random variables Y^{+} and Y^{-} proves the direct implication. The converse implication follows directly from Theorem 3.1 or by first taking $f(\cdot) = e^{it\cdot}$, $t \in \mathbb{R}$, in (3.27) and then, as previously done in the proof of Theorem 3.1, by solving a differential equation. $\square$

Remark 3.6 (i) If $m_0^{-} = 0$ and $m_0^{+} \neq 0$, Corollary 3.5 remains valid when replacing the identity (3.27) by

$$\mathbb{E}Xf(X) = m_0^{+}\mathbb{E}f(X + Y^{+}),$$

for all bounded Lipschitz functions f. *A similar proposition also holds true when* $m_0^- \neq 0$ *and* $m_0^+ = 0$.

(ii) When $X \sim ID(b, 0, \nu)$ *is nonnegative, then necessarily the support of its Lévy measure is in* $(0, +\infty)$, *the condition* $\int_0^1 u d\nu(u) < +\infty$ *is automatically satisfied and* $b_0 \geq 0$ *(see [84, Theorem 24.11]). In this context, for* $\mathbb{E}X < +\infty$, $b_0 = \mathbb{E}X - \int_0^{+\infty} u\nu(du)$, $m_0 = m_0^+ = \mathbb{E}X$ *and so, when* $\mathbb{E}X > 0$, *the characterizing identity (3.27) becomes*

$$\mathbb{E}Xf(X) = \mathbb{E}X\mathbb{E}f(X + Y), \tag{3.30}$$

where Y *is a random variable independent of* X *whose law is given by*

$$\mu_Y(du) = \frac{b_0}{\mathbb{E}X}\delta_0(du) + \frac{u\mathbb{1}_{(0,+\infty)}(u)}{\mathbb{E}X}\nu(du). \tag{3.31}$$

This agrees with the standard notion of size-bias distribution for finite mean nonnegative random variable (see, e.g., [7]) and recovers and extends a result there. The pair (Y^+, Y^-) *in (3.27) will be called the additive size-bias pair associated with* X.
(iii) For $X \geq 0$ *with finite first moment, there is a natural relationship between size-bias distribution and equilibrium distribution with respect to* X *(see [76]). Corollary 3.5 also leads to an extension of this relationship. Namely, for* X *as in the corollary and* f *bounded Lipschitz,*

$$\mathbb{E}f(X) - f(0) = \mathbb{E}Xf'(UX),$$
$$= m_0^+ \mathbb{E}f'(U(X + Y^+)) - m_0^- \mathbb{E}f'(U(X + Y^-)),$$

where U *is a uniform random variable on* $[0, 1]$ *independent of* X, Y^+ *and* Y^-.
(iv) For a nonnegative integer-valued random variable with finite mean X, *[82] introduces another distributional transformation in the following way: the random variable* X^{*r} *is said to have the* r-*equilibrium distribution if* $X^{*r} =_d U_{r,X^s}$, *where* $r > 0$, *where* X^s *has the size-bias distribution and where the random variable* $U_{r,n}$ *has the distribution of the number of white balls drawn in* $n - 1$ *draws in a standard Pólya urn scheme starting with* r *white balls and a single black ball. Now, let* $X \sim ID(b, 0, \nu)$ *be a nonnegative integer-valued random variable with finite first moment. Then, by (3.30), for all* f *bounded and Lipschitz*

$$\mathbb{E}Xf(X) = \mathbb{E}X\mathbb{E}f(X + Y),$$

where Y *is a random variable independent of* X *whose law is given by (3.31). Then, as in [82, Lemma 2.1], one shows that*

$$\mathbb{E}X\mathbb{E}f(X + Y) = \mathbb{E}X\mathbb{E}D^{(r)}f(X^{*r}),$$

with $X^{*r} =_d U_{r,X+Y}$ *and* $D^{(r)}(f)(k) = (k/r + 1)f(k + 1) - k/rf(k)$, *for all* $k \geq 1$, *and so*

$$\mathbb{E}Xf(X) = \mathbb{E}X\mathbb{E}D^{(r)}f(X^{*r}).$$

(v) Let us consider a nontrivial example for which the assumption (3.25) is not satisfied. Let X be a second-order Wiener chaos random variable such that, for all $k \geq 1$, $\lambda_k > 0$, with further $\sum_{k\geq 1}\lambda_k = +\infty$ (recall that $\sum_{k\geq 1}\lambda_k^2 < +\infty$). Then, for the Lévy measure ν_N given via

$$\nu_N(du) = \mathbb{1}_{(0,+\infty)}(u)\sum_{k=1}^{N}\frac{\exp\left(-\frac{u}{2\lambda_k}\right)}{2u}du,$$

we have

$$\int_{-1}^{+1}|u|\nu_N(du) = \sum_{k=1}^{N}\lambda_k\left(1-\exp\left(-\frac{1}{2\lambda_k}\right)\right) \geq \left(1-\exp\left(-\frac{1}{2\lambda_{\max}}\right)\right)\sum_{k=1}^{N}\lambda_k \xrightarrow[N\to+\infty]{} +\infty,$$

where $\lambda_{\max} = \max_{k\geq 1}\lambda_k$. So, by monotone convergence, $\int_{-1}^{+1}|u|\nu(du) = +\infty$. In particular, the Rosenblatt distribution belongs to this class of second Wiener chaos random variables, since asymptotically $\lambda_k \underset{k\to+\infty}{\sim} C_Dk^{D-1}$, for some $C_D > 0$ and $D \in (0, 1/2)$ (see Theorem 3.2 of [92]).

As seen in some of the examples presented above, characterizing identities can involve local operators, e.g., derivatives, while our generic characterization is non-local, involving difference operators. Let us explain, next, how to pass from one to the other also encouraging the reader to contemplate how this passage is linked to the notion of zero-bias distribution (see [47]) with an additive structure.

Remark 3.7 *As just indicated, let us present a general methodology valid for $X \sim ID(b, 0, \nu)$ such that $X \geq 0$ and $0 < \mathbb{E}X < +\infty$, to pass from the nonlocal characterization of Theorem 3.1 to a local characterization. Again, since $X \geq 0$, then necessarily the support of ν is in $(0, +\infty)$, $\int_0^1 u\nu(du) < +\infty$ and $b_0 \geq 0$ (see [84]). Hence, from the finite mean assumption and for all $v > 0$, $\eta(v) = \int_v^{+\infty}u\nu(du) < +\infty$. Therefore, denoting by μ the law of X, for any bounded Lipschitz function f,*

$$\begin{aligned}
Cov(X, f(X)) &= \mathbb{E}\int_0^{+\infty}(f(X+u) - f(X))u\nu(du)\\
&= \int_0^{+\infty}\int_0^{+\infty}\left(\int_0^u f'(x+v)dv\right)u\nu(du)\mu(dx)\\
&= \int_0^{+\infty}\int_0^{+\infty}f'(x+v)\eta(v)dv\mu(dx)\\
&= \int_0^{+\infty}f'(y)(\eta * \mu)(dy),
\end{aligned}\tag{3.32}$$

where $\eta * \mu$ is the convolution of the law μ with the positive Borel measure $\eta(dv) = \eta(v)dv$. Since $\eta * \mu$ is absolutely continuous with respect to the Lebesgue measure, denoting its Radon–Nikodym derivative by h, then $h(y) = \int_0^y \eta(y - v)\mu(dv)$, and (3.32) becomes

$$Cov(X, f(X)) = \int_0^{+\infty} f'(y)h(y)dy. \qquad (3.33)$$

In particular, when X has an exponential distribution, then $h(y) = ye^{-y}$ and (3.33) becomes the classical relation

$$Cov(X, f(X)) = \mathbb{E}Xf'(X). \qquad (3.34)$$

In general, $\eta * \mu$ is not a probability law, it is a positive measure, not necessarily finite, since $\int_0^{+\infty} \eta(v)dv = \int_0^{+\infty} u^2\nu(du)$. In case X is nondegenerate with $\mathbb{E}X^2 < +\infty$, i.e., $\int_1^{+\infty} u^2\nu(du) < +\infty$, (3.32) can be rewritten as

$$Cov(X, f(X)) = \eta((0, +\infty))\mathbb{E}f'(X + Y), \qquad (3.35)$$

where $\eta((0, +\infty)) = \int_0^{+\infty} u^2\nu(du) < +\infty$, and Y, with law $\eta/\eta((0, +\infty))$, is independent of X. In view of our previous corollary, it is a simple matter to modify the above arguments in case the condition $X \geq 0$ is not satisfied. The corresponding result is then given by the following proposition, whose proof is briefly sketched and whose statement is, again, also related to the notion of zero-bias distribution (see [47]).

Proposition 3.8 Let X be a nondegenerate random variable such that $\mathbb{E}X^2 < +\infty$. Let $b \in \mathbb{R}$, and let $\nu \neq 0$ be a Lévy measure such that

$$\int_{|u|>1} u^2\nu(du) < +\infty. \qquad (3.36)$$

Then,

$$Cov(X, f(X)) = \left(\int_{-\infty}^{+\infty} u^2\nu(du) \right) \mathbb{E}f'(X + Y), \qquad (3.37)$$

for all bounded Lipschitz functions f, where the random variables X and Y are independent, with the law of Y given by

$$\mu_Y(du) = \frac{\eta(u)}{\int_{-\infty}^{+\infty} u^2\nu(du)} du,$$

and where η is defined, for all $v \in \mathbb{R}$, by

$$\eta(v) := \eta_+(v)\mathbb{1}_{(0,+\infty)}(v) + \eta_-(v)\mathbb{1}_{(-\infty,0)}(v),$$

with η_+ and η_-, respectively, defined on $(0, +\infty)$ and on $(-\infty, 0)$ via

$$\eta_+(v) = \int_v^{+\infty} u\nu(du), \qquad \eta_-(v) = \int_{-\infty}^v (-u)\nu(du),$$

if and only if $X \sim ID(b, 0, \nu)$.

Proof Let us first sketch the proof of the direct implication. If μ denotes the law of X, then from Theorem 3.1 and our hypotheses,

$$
\begin{aligned}
Cov(X, f(X)) =& \mathbb{E} \int_{-\infty}^{+\infty} (f(X+u) - f(X))u\nu(du) \\
=& \int_{-\infty}^{+\infty} \int_0^{+\infty} (f(x+u) - f(x))u\nu(du)\mu(dx) \\
& + \int_{-\infty}^{+\infty} \int_{-\infty}^0 (f(x+u) - f(x))u\nu(du)\mu(dx) \\
=& \int_{-\infty}^{+\infty} \int_0^{+\infty} \left(\int_0^u f'(x+v)dv \right) u\nu(du)d\mu(dx) \\
& + \int_{-\infty}^{+\infty} \int_{-\infty}^0 \left(\int_u^0 f'(x+v)dv \right) (-u)\nu(du)\mu(dx) \\
=& \int_{-\infty}^{+\infty} \int_0^{+\infty} f'(x+v)\eta_+(v)dv\mu(dx) \\
& + \int_{-\infty}^{+\infty} \int_{-\infty}^0 f'(x+v)\eta_-(v)dvd\mu(dx) \\
=& \int_{-\infty}^{+\infty} \int_{-\infty}^{+\infty} f'(x+v)\Big(\eta_+(v)\mathbb{1}_{(0,+\infty)}(v) \\
& \qquad\qquad\qquad + \eta_-(v)\mathbb{1}_{(-\infty,0)}(v) \Big)dv\mu(dx) \\
=& \int_{-\infty}^{+\infty} \int_{-\infty}^{+\infty} f'(x+v)\eta(v)dv\mu(dx).
\end{aligned}
\tag{3.38}
$$

The conclusion then easily follows by the very definition of Y and the assumption (3.36). The converse implication is a direct consequence of the converse part of Theorem 3.1 or, as before, follows by taking $f(\cdot) = e^{it\cdot}, t \in \mathbb{R}$ in (3.37). $\qquad\square$

Remark 3.9 *(i) The previous proposition can, in particular, be applied to the two-sided exponential distribution with parameters $\alpha > 0$ and $\beta > 0$. In this case, the Lévy measure is given by $\nu(du) = \left(e^{-\alpha u}/u\mathbb{1}_{(0,+\infty)}(u) - e^{\beta u}/u\mathbb{1}_{(-\infty,0)}(u)\right) du$. Then, the condition (3.36) is readily satisfied, and the law of Y has the following density:*

$$f_Y(t) = \frac{\alpha^2\beta^2}{\alpha^2 + \beta^2} \left(\frac{e^{-\alpha t}}{\alpha}\mathbb{1}_{(0,+\infty)}(t) + \frac{e^{\beta t}}{\beta}\mathbb{1}_{(-\infty,0)}(t) \right). \tag{3.39}$$

(ii) As done in Corollary 3.5, Proposition 3.8 extends the notion of zero-bias distribution to all infinitely divisible nondegenerate distributions with finite variance. The random variable Y in (3.37) will be called the extended zero-bias distribution associated with X.

(iii) Another possible writing for (3.37), more in line with (3.27), is

$$Cov(X, f(X)) = \eta_+ ((0, +\infty)) \, \mathbb{E} f'(X + Y^+) + \eta_- ((-\infty, 0)) \, \mathbb{E} f'(X + Y^-),$$

$$(3.40)$$

where Y^+ and Y^- have respective law

$$\mu_{Y^+}(du) = \frac{\eta_+(u) \mathbb{1}_{(0, +\infty)}(u)}{\eta_+((0, +\infty))} du,$$

$$\mu_{Y^-}(du) = \frac{\eta_-(u) \mathbb{1}_{(-\infty, 0)}(u)}{\eta_-((-\infty, 0))} du,$$

and where

$$\eta_+ ((0, +\infty)) = \int_0^{+\infty} u^2 \nu(du), \quad \eta_- ((-\infty, 0)) := \int_{-\infty}^0 u^2 \nu(du).$$

Remark 3.10 *In the Stein's method literature, the size-bias and the zero-bias distributions are powerful tools which have been efficiently used in several situations. Classically, they have been used in conjunction with coupling techniques to produce quantitative results for Poisson and normal approximations (see, e.g., [31, 81]). More recently, these two concepts, combined again with coupling techniques, have also been used to prove concentration inequalities (see, e.g., [6, 15, 44, 46]). Note that Corollary 3.5 and Proposition 3.8 are characterizing results for nondegenerate infinitely divisible distributions whose Lévy measures satisfy appropriate moment conditions. These results suggest the introduction of extended notions of size-bias and zero-bias for nondegenerate random variables with finite first and second moments, respectively. For example, they could be of the type*

$$\mathbb{E} X f(X) = \mathbb{E} X^+ \mathbb{E} f(\tilde{X}^+) - \mathbb{E} X^- \mathbb{E} f(\tilde{X}^-),$$

for some random variables $\tilde{X}^+$, $\tilde{X}^-$ and would reduce to (3.27), i.e., to an additive framework, in the infinitely divisible case. In particular, thanks to these (covariance) representations, new methodologies to obtain concentration results for infinitely divisible distributions might be reachable and would, for example, complement [53, 54]. Finally, the reader is referred to [7] which emphasizes connections between size-bias distribution and several themes of probability theory.

It is important to note that the stable distributions with $\alpha \in (1, 2)$ do satisfy neither the assumptions of Corollary 3.5 nor those of Proposition 3.8. Nevertheless, our next result which is a mixture of the two previous ones characterizes infinitely divisible distributions with finite first moment, and in particular, the stable ones. For this purpose, we introduce the following functions, respectively, well defined on $(0, 1)$ and on $(-1, 0)$:

$$\eta_+(v) = \int_v^1 u\nu(du), \qquad \eta_-(v) = \int_{-1}^v (-u)\nu(du),$$

and note that since ν is a Lévy measure, for all $v \in (0, 1)$,

$$v\eta_+(v) \leq \int_v^1 u^2\nu(du) \leq \int_0^1 u^2\nu(du) < +\infty,$$

and similarly for η_-.

Proposition 3.11 *Let X be a nondegenerate random variable such that $\mathbb{E}|X| < +\infty$. Let $b \in \mathbb{R}$ and let ν be a Lévy measure such that*

$$0 < \int_{|u|>1} |u|\nu(du) < +\infty, \quad \text{and} \quad \int_{-1}^1 u^2\nu(du) > 0. \tag{3.41}$$

Then,

$$Cov(X, f(X)) = \left(\int_{-1}^1 u^2\nu(du)\right)\mathbb{E}f'(X + U) + m\mathbb{E}f(X + V_+) - m\mathbb{E}f(X + V_-),$$
$$\tag{3.42}$$

for all bounded Lipschitz functions f, where $m = m_+ + m_-$ with $m_\pm$ defined via

$$m_+ = \int_1^{+\infty} u\nu(du), \quad m_- = \int_{-\infty}^{-1} (-u)\nu(du),$$

and where the random variables X, U, V_+ and V_- are independent, with the laws of U, V_+ and V_-, respectively, given by

$$\mu_U(du) = \frac{\eta_+(u)\mathbb{1}_{(0,1)}(u) + \eta_-(u)\mathbb{1}_{(-1,0)}(u)}{\int_{-1}^{+1} u^2\nu(du)}du,$$

$$\mu_{V_+}(du) = \frac{m_-}{m}\delta_0(u) + \frac{u}{m}\mathbb{1}_{(1,+\infty)}(u)\nu(du),$$

$$\mu_{V_-}(du) = \frac{m_+}{m}\delta_0(u) + \frac{-u}{m}\mathbb{1}_{(-\infty,-1)}(u)\nu(du),$$

if and only if $X \sim ID(b, 0, \nu)$.

Proof First, let $X \sim ID(b, 0, \nu)$, and denote its law by μ. Then, from Theorem 3.1, for any bounded Lipschitz function,

$$Cov(X, f(X)) = \mathbb{E} \int_{-\infty}^{+\infty} (f(X + u) - f(X))u\nu(du)$$

$$= \mathbb{E} \int_{|u| \leq 1} (f(X + u) - f(X))u\nu(du)$$

$$+ \mathbb{E} \int_{|u| > 1} (f(X + u) - f(X))u\nu(du).$$

To continue, let us perform steps similar to those of Proposition 3.8 and Corollary 3.5 for, respectively, the first and second terms of the previous sum. For the first one,

$$\mathbb{E} \int_{|u| \leq 1} (f(X + u) - f(X))u\nu(du) = \mathbb{E} \int_0^1 (f(X + u) - f(X))u\nu(du)$$

$$+ \mathbb{E} \int_{-1}^0 (f(X + u) - f(X))u\nu(du)$$

$$= \int_{-\infty}^{+\infty} \int_0^1 \left(\int_0^u f'(x + v)dv \right) u\nu(du)d\mu(dx)$$

$$+ \int_{-\infty}^{+\infty} \int_{-1}^0 \left(\int_u^0 f'(x + v)dv \right) (-u)\nu(du)\mu(dx)$$

$$= \int_{-\infty}^{+\infty} \int_0^1 f'(x + v)\eta_+(v)dv\mu(dx)$$

$$+ \int_{-\infty}^{+\infty} \int_{-1}^0 f'(x + v)\eta_-(v)dvd\mu(dx)$$

$$= \int_{-\infty}^{+\infty} \int_{-\infty}^{+\infty} f'(x + v)\Big(\eta_+(v)\mathbb{1}_{(0,1)}(v)$$

$$+ \eta_-(v)\mathbb{1}_{(-1,0)}(v) \Big)dv\mu(dx)$$

$$= \int_{-\infty}^{+\infty} \int_{-\infty}^{+\infty} f'(x + v)\eta(v)dv\mu(dx).$$

For the second term,

$$\mathbb{E} \int_{|u| > 1} (f(X + u) - f(X))u\nu(du) = \mathbb{E} \int_{|u| > 1} f(X + u)u\nu(du) - \mathbb{E}f(X) \int_{|u| > 1} u\nu(du)$$

$$= \mathbb{E} \int_1^{+\infty} f(X + u)u\nu(du) - \mathbb{E} \int_{-\infty}^{-1} f(X + u)(-u)\nu(du)$$

$$- \mathbb{E}f(X) \int_{|u| > 1} u\nu(du)$$

$$= m \left(\mathbb{E}f(X + V_+) - \mathbb{E}f(X + V_-) \right).$$

The conclusion then easily follows from the very definition of U, V_+ and V_- and the assumption (3.41). The converse implication is a direct consequence of the converse part of Theorem 3.1 or, as before, follows by taking $f(\cdot) = e^{it\cdot}$, $t \in \mathbb{R}$ in (3.42). $\square$

Chapter 4
General Upper Bounds by Fourier Methods

The Fourier methodology developed in [91] to study the Stein's equation in the Gaussian setting, often nowadays referred to as the Stein–Tikhomirov method, has been extended in [4] to provide rates of convergence in Kolmogorov or in smooth Wasserstein distance for sequences $(X_n)_{n \geq 1}$ converging toward X_∞. This approach leads to quantitative estimates when X_∞ is a second-order Wiener chaos, or the generalized Dickman distribution or even the symmetric α-stable one. Corollary 3.5, or Proposition 3.8, or even the stable characterizing identities of the previous chapter allow extensions of the aforementioned estimates to classes of infinitely divisible sequences. The forthcoming results are general and have a non-empty intersection with those on the Dickman distribution presented in [4].

Theorem 4.1 *Let $X_n \sim ID(b_n, 0, \nu_n)$, $n \geq 1$, be a sequence of nondegenerate random variables converging in law toward the nondegenerate $X_\infty \sim ID(b_\infty, 0, \nu_\infty)$, with also $\mathbb{E}|X_n| < +\infty$, $\mathbb{E}|X_\infty| < +\infty$ and*

$$\int_{-1}^{+1} |u| \nu_n(du) < +\infty, \qquad \int_{-1}^{+1} |u| \nu_\infty(du) < +\infty, \tag{4.1}$$

$n \geq 1$. Further, for all $t \in \mathbb{R}$, let

$$|\varphi_\infty(t)| \int_0^{|t|} \frac{ds}{|\varphi_\infty(s)|} \leq C_\infty |t|^{p_\infty}, \tag{4.2}$$

where φ_∞ is the characteristic function of X_∞ and where $C_\infty > 0$, $p_\infty \geq 1$. Let the law of X_∞ be absolutely continuous with respect to the Lebesgue measure and have a bounded density. Then,

$$d_K(X_n, X_\infty) \leq C'_\infty \Delta_n^{\frac{1}{p_\infty + 2}},$$

© The Author(s), under exclusive license to Springer Nature Switzerland AG 2019
B. Arras and C. Houdré, *On Stein's Method for Infinitely Divisible Laws with Finite First Moment*, SpringerBriefs in Probability and Mathematical Statistics, https://doi.org/10.1007/978-3-030-15017-4_4

where

$$\Delta_n = |(m_0^n)^+ - (m_0^\infty)^+| + |(m_0^n)^- - (m_0^\infty)^-|$$
$$+ (m_0^\infty)^+ \mathbb{E}|Y_n^+ - Y_\infty^+| + (m_0^\infty)^- \mathbb{E}|Y_n^- - Y_\infty^-|,$$

where $(m_0^n)^\pm$, $Y_n^\pm$ and $(m_0^\infty)^\pm$, $Y_\infty^\pm$ are the quantities defined in Corollary 3.5, respectively, associated with X_n and X_∞, and where $C_\infty' > 0$ depends on the supremum norm of the density of X_∞ but is independent of n.

Proof From Corollary 3.5 applied to X_n and X_∞, let

$$\Delta_n^\pm(t) := (m_0^n)^\pm \varphi_{Y_n^\pm}(t) - (m_0^\infty)^\pm \varphi_{Y_\infty^\pm}(t),$$
$$S_\infty(t) := (m_0^\infty)^+ \varphi_{Y_\infty^+}(t) - (m_0^\infty)^- \varphi_{Y_\infty^-}(t),$$
$$\varepsilon_n(t) := \varphi_n(t) - \varphi_\infty(t),$$

where $\varphi_{Y_n^\pm}$ and $\varphi_{Y_\infty^\pm}$ are the characteristic functions of $Y_n^\pm$ and $Y_\infty^\pm$. Now, thanks to the identity (3.27) applied to the test functions $f(\cdot) = e^{it\cdot}$,

$$\frac{1}{i}\frac{d}{dt}(\varphi_\infty(t)) = \varphi_\infty(t)S_\infty(t),$$
$$\frac{1}{i}\frac{d}{dt}(\varphi_n(t)) = \varphi_n(t)S_\infty(t) + \varphi_n(t)\Delta_n^+(t) - \varphi_n(t)\Delta_n^-(t).$$

Subtracting these last two expressions, recalling also that the characteristic function of an ID law never vanishes, leads to:

$$\frac{d}{dt}(\varepsilon_n(t)) = \frac{\varepsilon_n(t)}{\varphi_\infty(t)}\frac{d}{dt}(\varphi_\infty(t)) + i\varphi_n(t)(\Delta_n^+(t) - \Delta_n^-(t)),$$

since

$$S_\infty(t) = \frac{1}{i\varphi_\infty(t)}\frac{d}{dt}(\varphi_\infty(t)).$$

Then, straightforward computations imply that for all $t \geq 0$:

$$\varepsilon_n(t) = i\varphi_\infty(t)\int_0^t \frac{\varphi_n(s)}{\varphi_\infty(s)}\big((\Delta_n^+(s) - \Delta_n^-(s))\big)ds,$$

and similarly for $t \leq 0$. Let us next bound the difference, $\Delta_n^+(s) - \Delta_n^-(s)$. First,

$$|\Delta_n^+(s) - \Delta_n^-(s)| \leq I + II + III + IV,$$

where

$$I := |(m_0^n)^+ - (m_0^\infty)^+|,$$
$$II := |(m_0^n)^- - (m_0^\infty)^-|,$$
$$III := (m_0^\infty)^+ |\varphi_{Y_n^+}(s) - \varphi_{Y_\infty^+}(s)| \le (m_0^\infty)^+ |s| \mathbb{E}|Y_n^+ - Y_\infty^+|,$$
$$IV := (m_0^\infty)^- |\varphi_{Y_n^-}(s) - \varphi_{Y_\infty^-}(s)| \le (m_0^\infty)^- |s| \mathbb{E}|Y_n^- - Y_\infty^-|.$$

Hence,

$$|\varepsilon_n(t)| \le |\varphi_\infty(t)| \int_0^t \frac{ds}{|\varphi_\infty(s)|} \left(|(m_0^n)^+ - (m_0^\infty)^+| + |(m_0^n)^- - (m_0^\infty)^-| \right)$$
$$+ |\varphi_\infty(t)| \int_0^t \frac{|s|ds}{|\varphi_\infty(s)|} \left((m_0^\infty)^+ \mathbb{E}|Y_n^+ - Y_\infty^+| + (m_0^\infty)^- \mathbb{E}|Y_n^- - Y_\infty^-| \right).$$

Then, using (4.2), together with the definition of Δ_n, entails

$$|\varepsilon_n(t)| \le C_\infty (|t|^{p_\infty} + |t|^{p_\infty+1}) \Delta_n. \tag{4.3}$$

Since the law of X_∞ has a bounded density, applying the classical Esseen inequality (see, e.g., [77, Theorem 5.1]) gives, for all $T > 0$,

$$d_K(X_n, X_\infty) \le C_1 \int_{-T}^T \frac{|\varepsilon_n(t)|}{|t|} dt + C_2 \frac{\|h_\infty\|_\infty}{T}, \tag{4.4}$$

where C_1 and C_2 are positive (absolute) constants, while $\|h_\infty\|_\infty$ is the essential supremum of the density h_∞ of the law of X_∞. Next, plugging (4.3) into (4.4), it follows that

$$d_K(X_n, X_\infty) \le C_1' \left(T^{p_\infty} + T^{p_\infty+1} \right) \Delta_n + \frac{C_2'}{T}.$$

The choice $T = (1/\Delta_n)^{\frac{1}{p_\infty+2}}$ concludes the proof. $\qquad\square$

Remark 4.2 *(i) Let us briefly discuss the growth condition on the limiting characteristic function, namely, the requirement that for all $t \in \mathbb{R}$,*

$$L(\varphi)(t) := |\varphi(t)| \int_0^{|t|} \frac{ds}{|\varphi(s)|} \le C|t|^p, \tag{4.5}$$

for some $C > 0$ and $p \ge 1$. When the limiting distribution is the standard normal one, the functional $L(\varphi)$ is the Dawson integral associated with the normal. It decreases to zero at infinity, and for any $t \in \mathbb{R}$,

$$L(\varphi)(t) := e^{-\frac{t^2}{2}} \int_0^{|t|} e^{\frac{s^2}{2}} ds \le \frac{2|t|}{1+t^2}. \tag{4.6}$$

Various behaviors are possible for this (generalized Dawson) functional, see [4]. As detailed below, in a general gamma setting, (4.5) holds true with $p = 1$ while in the stable case (see Lemma 10 of Appendix B in [4]), for $1 < \alpha < 2$, and $t > 0$,

$$L(\varphi)(t) \le \left(t^{1-\alpha}/c + Ce^{-ct^\alpha} \right), \tag{4.7}$$

where $C = \int_0^{c^{-1/\alpha}} e^{cs^\alpha} ds$ and $c = c_1 + c_2$ as given in (2.10). In particular, $C \le e/c^{1/\alpha}$. Moreover, for t small, (4.7) can be replaced by

$$L(\varphi)(t) \le |t|.$$

Then, for some constant $C' > 0$ only depending on α and c, and for all $t \in \mathbb{R}$,

$$L(\varphi)(t) \le C' \frac{|t|}{1 + |t|^\alpha}. \tag{4.8}$$

For the compound Poisson case with finite Lévy measure ν,

$$L\varphi(t) = e^{\nu(\mathbb{R}) \int_{-\infty}^{+\infty} (\cos(ut)-1)\nu_0(du)} \int_0^{|t|} e^{\nu(\mathbb{R}) \int_{-\infty}^{+\infty} (1-\cos(us))\nu_0(du)} ds \le e^{2\nu(\mathbb{R})} |t|, \quad t \in \mathbb{R}.$$

For the generalized Dickman distribution as considered in [4], a linear growth can also be obtained from the corresponding characteristic function.
(ii) As it is well known, e.g., see [81],

$$d_K(X_n, X_\infty) \le \sqrt{2\|h_\infty\|_\infty W_1(X_n, X_\infty)}, \tag{4.9}$$

where again $\|h_\infty\|_\infty$ is the supremum norm of h_∞, the bounded density of the law of X_∞, and where W_1 is the Wasserstein-1 distance as given in (2.17) which also admits the following well-known representation:

$$W_1(X, Y) = \sup_{h \in Lip(1)} |\mathbb{E}h(X) - \mathbb{E}h(Y)|,$$

for X, Y random variables with finite first moment. Therefore, to go beyond the bounded density case, e.g., to consider discrete limiting laws, it is natural to explore convergence rates in (smooth) Wasserstein. Under uniform (exponential) integrability, such issues can be tackled. For example, instead of the bounded density assumption, let, for some $\lambda > 0$ and $\alpha \in (0, 1]$,

$$\sup_{n\ge1} \mathbb{E}\, e^{\lambda|X_n|^\alpha} < +\infty, \quad , i.e., \quad \sup_{n\ge1} \int_{|u|>1} e^{\lambda|u|^\alpha} \nu_n(du) < +\infty, \tag{4.10}$$

then,

$$dw_{p_\infty+2}(X_n, X_\infty) \leq C'_\infty \Delta_n |\ln \Delta_n|^{\frac{1}{2\alpha}}, \tag{4.11}$$

where C'_∞ and Δ_n are as in the previous theorem. The proof of (4.11) uses the pointwise estimate (4.3) combined with the assumption (4.10) and with the statement and conclusion of [4, Theorem 1].

Proposition 3.8 also provides quantitative upper bounds in Kolmogorov distance. This is the content of the next proposition whose statement is similar to that of Theorem 4.1.

Proposition 4.3 *Let $X_n \sim ID(b_n, 0, \nu_n)$, $n \geq 1$, be a sequence of nondegenerate random variables converging in law toward $X_\infty \sim ID(b_\infty, 0, \nu_\infty)$ (nondegenerate) and such that $\mathbb{E}|X_n|^2 < +\infty$, $\mathbb{E}|X_\infty|^2 < +\infty$, $n \geq 1$. Let also, for all $t \in \mathbb{R}$,*

$$|\varphi_\infty(t)| \int_0^{|t|} \frac{ds}{|\varphi_\infty(s)|} \leq C_\infty |t|^{p_\infty}, \tag{4.12}$$

where φ_∞ is the characteristic function of X_∞ and where $C_\infty > 0$, $p_\infty \geq 1$. Let the law of X_∞ be absolutely continuous with respect to the Lebesgue measure and have a bounded density, then

$$d_K(X_n, X_\infty) \leq C'_\infty \Delta_n^{\frac{1}{p_\infty+3}}, \tag{4.13}$$

where

$$\Delta_n = |\eta_n - \eta_\infty| + |\mathbb{E}X_n - \mathbb{E}X_\infty| + \mathbb{E}|Y_n - Y_\infty|, \tag{4.14}$$

where Y_n and Y_∞ are the random variables defined in Proposition 3.8, respectively, associated with X_n and X_∞, where

$$\eta_n := \int_{-\infty}^{+\infty} u^2 \nu_n(du), \qquad \eta_\infty := \int_{-\infty}^{+\infty} u^2 \nu_\infty(du),$$

and where $C'_\infty > 0$ is independent of n.

Proof The proof of this proposition is very similar to the proof of Theorem 4.1 and so it is only sketched. From Proposition 3.8 applied to X_n and X_∞, let

$$m_n := \mathbb{E}X_n, \qquad m_\infty := \mathbb{E}X_\infty,$$
$$\Delta_n(t) := t(\eta_\infty \varphi_{Y_\infty}(t) - \eta_n \varphi_{Y_n}(t)) + i(m_n - m_\infty),$$
$$R_\infty(t) := -t\eta_\infty \varphi_{Y_\infty}(t) + im_\infty,$$
$$\varepsilon(t) := \varphi_n(t) - \varphi_\infty(t).$$

Next, thanks to the identity (3.37) applied to the test functions $f(\cdot) = e^{it\cdot}$,

$$\frac{d}{dt}(\varphi_n(t)) = R_\infty(t)\varphi_n(t) + \varphi_n(t)\Delta_n(t),$$

$$\frac{d}{dt}(\varphi_\infty(t)) = R_\infty(t)\varphi_\infty(t).$$

Subtracting the last two expressions, it follows that

$$\frac{d}{dt}(\varepsilon_n(t)) = \frac{\varepsilon_n(t)}{\varphi_\infty(t)}\frac{d}{dt}(\varphi_\infty(t)) + \varphi_n(t)\Delta_n(t).$$

Then, straightforward computations imply that for all $t \geq 0$,

$$\varepsilon_n(t) = \varphi_\infty(t)\int_0^t \frac{\varphi_n(s)}{\varphi_\infty(s)}\Delta_n(s)ds,$$

and similarly for $t \leq 0$. Combining this last expression with (4.12) and with standard estimates give

$$|\varepsilon_n(t)| \leq C_\infty(|t|^{p_\infty} + |t|^{p_\infty+1} + |t|^{p_\infty+2})\Delta_n.$$

Finally, proceeding as in the end of the proof of Theorem 4.1 concludes the proof.   □

Remark 4.4 *(i) Since the random variables $(Y_n^\pm, Y_\infty^\pm)$ (resp. (Y_n, Y_∞)) in Theorem 4.1 (resp. Proposition 4.3) are independent of (X_n, X_∞), one can choose any of their couplings. In particular, Δ_n in Theorem 4.1 can be replaced by*

$$\Delta_n = |(m_0^n)^+ - (m_0^\infty)^+| + |(m_0^n)^- - (m_0^\infty)^-|$$
$$+ (m_0^\infty)^+ W_1(Y_n^+, Y_\infty^+) + (m_0^\infty)^- W_1(Y_n^-, Y_\infty^-). \qquad (4.15)$$

Similarly, the quantity Δ_n of Proposition 4.3 can be replaced by

$$\Delta_n = |\eta_n - \eta_\infty| + |\mathbb{E}X_n - \mathbb{E}X_\infty| + W_1(Y_n, Y_\infty). \qquad (4.16)$$

(ii) Recall that the Wasserstein-1 distance between two random variables X and $\tilde{X}$ both having finite first moment, and respective law μ and $\tilde{\mu}$ can also be represented as

$$W_1(X, \tilde{X}) = \int_{-\infty}^{+\infty} |F_\mu(t) - F_{\tilde{\mu}}(t)|dt, \qquad (4.17)$$

where F_μ and $F_{\tilde{\mu}}$ are the respective cumulative distribution functions of μ and $\tilde{\mu}$. Combining the above with Proposition 3.8, (4.16) becomes

$$\Delta_n = |\eta_n - \eta_\infty| + |\mathbb{E}X_n - \mathbb{E}X_\infty| + \int_{-\infty}^0 \left| \int_{-\infty}^t (-v)(t-v) \left(\frac{\nu_n(dv)}{\eta_n} - \frac{\nu_\infty(dv)}{\eta_\infty} \right) \right| dt$$

$$+ \int_0^{+\infty} \left| \int_0^{+\infty} v(v \wedge t) \left(\frac{\nu_n(dv)}{\eta_n} - \frac{\nu_\infty(dv)}{\eta_\infty} \right) + \int_{-\infty}^0 v^2 \left(\frac{\nu_n(dv)}{\eta_n} - \frac{\nu_\infty(dv)}{\eta_\infty} \right) \right| dt.$$

$$(4.18)$$

(iii) Next, for the second-order chaoses $X_n = \sum_{k=1}^{+\infty} \lambda_{n,k}(Z_k^2 - 1)/2$, $n \geq 1$ *and*
$X_\infty = \sum_{k=1}^{+\infty} \lambda_{\infty,k}(Z_k^2 - 1)/2$, *with* $\lambda_{n,k} > 0$ *and* $\lambda_{\infty,k} > 0$, *for all* $k \geq 1$, Δ_n *in*
(4.18) becomes

$$\Delta_n = 2 \int_0^{+\infty} \left| \sum_{k=1}^{+\infty} (\lambda_{\infty,k}^2 e^{-\frac{t}{2\lambda_{\infty,k}}} - \lambda_{n,k}^2 e^{-\frac{t}{2\lambda_{n,k}}}) \right| dt. \qquad (4.19)$$

Similar computations can be done using (4.15) and (4.17).

(iv) Again, the Kolmogorov distance can be replaced by a smooth Wasserstein one.
(Replacing also the bounded density assumption.) Indeed, if for some $\lambda > 0$ *and*
$\alpha \in (0, 1]$, $\sup_{n \geq 1} \mathbb{E} \, e^{\lambda |X_n|^\alpha} < +\infty$, *then*

$$d_{W_{p_\infty + 3}}(X_n, X_\infty) \leq C_\infty' \Delta_n |\ln \Delta_n|^{\frac{1}{2\alpha}},$$

as easily seen by simple modifications of the techniques presented above.
     (v) Any sequence of infinitely divisible random variables converging in law has
a limiting distribution which is itself infinitely divisible, e.g., [84, Lemma 7.8]. It is
thus natural to ask for conditions for such convergence as well as for quantitative
versions of it. On this subject, [84, Theorem 8.7] provides necessary and sufficient
conditions ensuring the weak convergence of sequences of infinitely divisible distri-
butions. Namely, it requires that, as $n \to +\infty$,

$$\beta_n = b_n + \int_{-\infty}^{+\infty} u \left(c(u) - \mathbb{1}_{|u| \leq 1} \right) \nu_n(du) \longrightarrow \beta_\infty,$$

and that

$$u^2 c(u) d\nu_n \Longrightarrow u^2 c(u) d\nu_\infty,$$

for some bounded continuous function c from $\mathbb{R}$ *to* $\mathbb{R}$ *such that* $c(u) = 1 + o(|u|)$,
as $|u| \to 0$, *and* $c(u) = O(1/|u|)$, *as* $|u| \to +\infty$. *Therefore, Theorem 4.1 and*
Proposition 4.3 provide quantitative versions of these results.

The previous results do not encompass the case of the stable distributions since
neither (4.1) nor $\mathbb{E}|X_\infty|^2 < +\infty$ are satisfied. To obtain quantitative convergence
results toward more general ID distributions, let us present a result valid for some

classes of self-decomposable laws. *Again, below and elsewhere, we follow [84] and use the terminology increasing or decreasing in a non-strict sense.*

Proposition 4.5 *Let $X_n \sim ID(b_n, 0, \nu_n)$, $n \geq 1$, be a sequence of nondegenerate random variables converging in law toward $X_\infty \sim ID(b_\infty, 0, \nu_\infty)$ (nondegenerate) and such that $\mathbb{E}|X_n| < +\infty$, $n \geq 1$, $\mathbb{E}|X_\infty| < +\infty$, and let*

$$\nu_n(du) := \frac{\psi_{1,n}(u)}{u} \mathbb{1}_{(0,+\infty)}(u)du + \frac{\psi_{2,n}(-u)}{(-u)} \mathbb{1}_{(-\infty,0)}(u)du,$$

$$\nu_\infty(du) := \frac{\psi_{1,\infty}(u)}{u} \mathbb{1}_{(0,+\infty)}(u)du + \frac{\psi_{2,\infty}(-u)}{(-u)} \mathbb{1}_{(-\infty,0)}(u)du,$$

where $\psi_{1,n}$, $\psi_{2,n}$, $\psi_{1,\infty}$ and $\psi_{2,\infty}$ are nonnegative decreasing functions on $(0, +\infty)$. Let also, for all $t \in \mathbb{R}$,

$$|\varphi_\infty(t)| \int_0^{|t|} \frac{ds}{|\varphi_\infty(s)|} \leq C_\infty |t|^{p_\infty}, \tag{4.20}$$

where φ_∞ is the characteristic function of X_∞ and where $C_\infty > 0$, $p_\infty \geq 1$. Finally, let the law of X_∞ be absolutely continuous with respect to the Lebesgue measure and have a bounded density. Then,

$$d_K(X_n, X_\infty) \leq C'_\infty (\Delta_n)^{\frac{1}{p_\infty+2}},$$

where

$$\Delta_n = |\mathbb{E}X_n - \mathbb{E}X_\infty| + \int_0^1 |u||\psi_{1,n}(u) - \psi_{1,\infty}(u)|du + \int_1^{+\infty} |\psi_{1,n}(u) - \psi_{1,\infty}(u)|du$$

$$+ \int_{-1}^0 |u||\psi_{2,n}(-u) - \psi_{2,\infty}(-u)|du + \int_{-\infty}^{-1} |\psi_{2,n}(-u) - \psi_{2,\infty}(-u)|du,$$

and where $C'_\infty > 0$ is independent of n.

Proof Again, this proof is very similar to the proof of Theorem 4.1 and so it is only sketched. Let

$$m_n := \mathbb{E}X_n, \qquad m_\infty := \mathbb{E}X_\infty,$$

$$\Delta_n(t) := m_n - m_\infty + \int_{-\infty}^{+\infty} (e^{itu} - 1)u\nu_n(du) - \int_{-\infty}^{+\infty} (e^{itu} - 1)u\nu_\infty(du),$$

$$S_\infty(t) := m_\infty + \int_{-\infty}^{+\infty} (e^{itu} - 1)u\nu_\infty(du),$$

$$\varepsilon_n(t) := \varphi_n(t) - \varphi_\infty(t).$$

Applying the identity (3.7) to X_n and X_∞ with $f(\cdot) = e^{it\cdot}$ gives

$$\frac{d}{dt}(\varphi_n(t)) = i\Delta_n(t)\varphi_n(t) + iS_\infty(t)\varphi_n(t),$$
$$\frac{d}{dt}(\varphi_\infty(t)) = iS_\infty(t)\varphi_\infty(t),$$

and thus,

$$\frac{d}{dt}(\varepsilon_n(t)) = \frac{\varphi'_\infty(t)}{\varphi_\infty(t)}\varepsilon_n(t) + i\Delta_n(t)\varphi_n(t).$$

Therefore, for all $t \geq 0$,

$$\varepsilon_n(t) = i\varphi_\infty(t)\int_0^t \frac{\varphi_n(s)}{\varphi_\infty(s)}\Delta_n(s)ds,$$

and similarly for $t \leq 0$. Let us now bound the quantity $\Delta_n(\cdot)$

$$|\Delta_n(s)| \leq |m_n - m_\infty| + \left|\int_{-\infty}^{+\infty}(e^{isu} - 1)u\nu_n(du) - \int_{-\infty}^{+\infty}(e^{isu} - 1)u\nu_\infty(du)\right|$$

$$\leq 2(1 + |s|)\left(|m_n - m_\infty| + + \int_0^1 |u||\psi_{1,n}(u) - \psi_{1,\infty}(u)|du\right.$$

$$+ \int_1^{+\infty}|\psi_{1,n}(u) - \psi_{1,\infty}(u)|du + \int_{-1}^0 |u||\psi_{2,n}(-u) - \psi_{2,\infty}(-u)|du$$

$$\left.+ \int_{-\infty}^{-1}|\psi_{2,n}(-u) - \psi_{2,\infty}(-u)|du\right),$$

$$\leq 2(1 + |s|)\Delta_n.$$

This implies

$$|\varepsilon_n(t)| \leq C_\infty(|t|^{p_\infty} + |t|^{p_\infty+1})\Delta_n.$$

To conclude the proof, proceed as in the end of the proof of Theorem 4.1.   □

Remark 4.6 *Recalling (4.8), note that the stable distributions do satisfy the assumptions of Proposition 4.5. However, the very specific properties of their Lévy measure entail detailed computations in order to reach a precise rate of convergence. To illustrate how our methodology can be adapted to obtain a rate of convergence toward a stable law, we present an example pertaining to the domain of normal attraction of the symmetric α-stable distribution. Let $1 < \alpha < 2$ and let $c := (1 - \alpha)/(2\Gamma(2 - \alpha)\cos(\alpha\pi/2))$ and $\lambda := (2c)^{1/\alpha}$. Then, denote by $f_1(x) := \alpha(2\lambda)^{-1}(1 + |x|/\lambda)^{-\alpha-1}$ the density of the Pareto law with parameters $\alpha > 0$ and $\lambda > 0$. As well known, this random variable is infinitely divisible, see [87, Chap. IV,*

Example 11.6], and belongs to the domain of normal attraction of the symmetric α-stable distribution, [77]. Our version of the Pareto density differs from the one considered in [33, 60, 94] given by $f(x) := \alpha\lambda^\alpha/(2|x|^{\alpha+1})\mathbb{1}_{|x|>\lambda}$, which is not infinitely divisible. Indeed, since f is symmetric, its characteristic function φ is real-valued, and by standard computations,

$$\varphi(s) = 1 - s^\alpha + \alpha \int_0^1 \frac{1 - \cos(\lambda y s)}{y^{\alpha+1}} dy \leq 1 - s^\alpha + \frac{\alpha s^2 \lambda^2}{2(2 - \alpha)},$$

$s > 0$. Now, it is not difficult to see that the above right-hand side can take negative values, e.g., for $\alpha = 3/2$ and $s = 2$, so φ would then have to vanish, contradicting infinite divisibility.

Proposition 4.7 *Let $(\xi_i)_{i\geq 1}$ be a sequence of iid random variables such that $\xi_1 \sim f_1$. For $1 < \alpha < 2$, let $X_n = \sum_{i=1}^n \xi_i/n^{1/\alpha}$, $n \geq 1$, and let $X_\infty \sim S\alpha S$ have characteristic function $\varphi_\infty(t) = \exp(-|t|^\alpha)$, $t \in \mathbb{R}$. Then,*

$$d_K(X_n, X_\infty) \leq \frac{C}{n^{\frac{2}{\alpha}-1}}, \tag{4.21}$$

for some $C > 0$ which depends only on α.

Proof Let φ_n be the characteristic function of X_n, $n \geq 1$. Adopting the notation of the proof of Proposition 4.5, for all $t \geq 0$,

$$\varepsilon_n(t) = i\varphi_\infty(t) \int_0^t \frac{\varphi_n(s)}{\varphi_\infty(s)} \Delta_n(s) ds,$$

and similarly for $t \leq 0$. But thanks to the identity (3.7) applied to X_n and X_∞ with $f(\cdot) = e^{is\cdot}$,

$$i\Delta_n(s) = \frac{\varphi_n'(s)}{\varphi_n(s)} - \frac{\varphi_\infty'(s)}{\varphi_\infty(s)}.$$

Moreover, for $s > 0$,

$$\frac{\varphi_\infty'(s)}{\varphi_\infty(s)} = -\alpha s^{\alpha-1}.$$

By standard computations, for $s > 0$,

$$\varphi_1(s) = \alpha \int_1^{+\infty} \cos(\lambda s(y - 1)) \frac{dy}{y^{\alpha+1}}$$

$$= \cos(\lambda s)\left(1 - s^\alpha + \int_0^1 \frac{1 - \cos(\lambda s y)}{y^{\alpha+1}} \alpha dy\right) + \sin(\lambda s)\left(\int_1^{+\infty} \frac{\sin(\lambda s z)}{z^{\alpha+1}} \alpha dz\right)$$

$$= 1 - s^\alpha + \psi_1(s) + \psi_2(s) + \psi_3(s),$$

where

$$\psi_1(s) = \int_0^1 \frac{1 - \cos(\lambda s y)}{y^{\alpha+1}} \alpha dy,$$

$$\psi_2(s) = (\cos(\lambda s) - 1)\left(\int_1^{+\infty} \frac{\cos(\lambda s z)}{z^{\alpha+1}} \alpha dz \right),$$

and

$$\psi_3(s) = \sin(\lambda s)\left(\int_1^{+\infty} \frac{\sin(\lambda s z)}{z^{\alpha+1}} \alpha dz \right).$$

Then,

$$\frac{\varphi_n'(s)}{\varphi_n(s)} = n^{1-\frac{1}{\alpha}} \frac{\varphi_1'\left(\frac{s}{n^{\frac{1}{\alpha}}}\right)}{\varphi_1\left(\frac{s}{n^{\frac{1}{\alpha}}}\right)}$$

$$= \frac{-\alpha s^{\alpha-1} + n^{1-\frac{1}{\alpha}}\psi_1'\left(\frac{s}{n^{\frac{1}{\alpha}}}\right) + n^{1-\frac{1}{\alpha}}\psi_2'\left(\frac{s}{n^{\frac{1}{\alpha}}}\right) + n^{1-\frac{1}{\alpha}}\psi_3'\left(\frac{s}{n^{\frac{1}{\alpha}}}\right)}{1 - \frac{s^\alpha}{n} + \psi_1\left(\frac{s}{n^{\frac{1}{\alpha}}}\right) + \psi_2\left(\frac{s}{n^{\frac{1}{\alpha}}}\right) + \psi_3\left(\frac{s}{n^{\frac{1}{\alpha}}}\right)},$$

implying that

$$\Delta_n(s) = \frac{-\alpha s^{\alpha-1} + n^{1-\frac{1}{\alpha}}\psi_1'\left(\frac{s}{n^{\frac{1}{\alpha}}}\right) + n^{1-\frac{1}{\alpha}}\psi_2'\left(\frac{s}{n^{\frac{1}{\alpha}}}\right) + n^{1-\frac{1}{\alpha}}\psi_3'\left(\frac{s}{n^{\frac{1}{\alpha}}}\right)}{1 - \frac{s^\alpha}{n} + \psi_1\left(\frac{s}{n^{\frac{1}{\alpha}}}\right) + \psi_2\left(\frac{s}{n^{\frac{1}{\alpha}}}\right) + \psi_3\left(\frac{s}{n^{\frac{1}{\alpha}}}\right)} + \alpha s^{\alpha-1}$$

$$= \frac{n^{1-\frac{1}{\alpha}}\psi_1'\left(\frac{s}{n^{\frac{1}{\alpha}}}\right) + n^{1-\frac{1}{\alpha}}\psi_2'\left(\frac{s}{n^{\frac{1}{\alpha}}}\right) + n^{1-\frac{1}{\alpha}}\psi_3'\left(\frac{s}{n^{\frac{1}{\alpha}}}\right) - \alpha\frac{s^{2\alpha-1}}{n} + \alpha s^{\alpha-1}\left(\psi_1\left(\frac{s}{n^{\frac{1}{\alpha}}}\right) + \psi_2\left(\frac{s}{n^{\frac{1}{\alpha}}}\right) + \psi_3\left(\frac{s}{n^{\frac{1}{\alpha}}}\right)\right)}{1 - \frac{s^\alpha}{n} + \psi_1\left(\frac{s}{n^{\frac{1}{\alpha}}}\right) + \psi_2\left(\frac{s}{n^{\frac{1}{\alpha}}}\right) + \psi_3\left(\frac{s}{n^{\frac{1}{\alpha}}}\right)}.$$

Before, bounding the quantity ε_n let us provide bounds on the functions ψ_1, ψ_2, and ψ_3 and their derivatives. For $s > 0$,

$$|\psi_1(s)| \leq C_1 s^2, \quad |\psi_2(s)| \leq C_2 s^2, \quad |\psi_3(s)| \leq C_3 s^2, \tag{4.22}$$

and

$$|\psi_1'(s)| \leq C_4 s, \quad |\psi_2'(s)| \leq C_5(s + s^2), \quad |\psi_3'(s)| \leq C_6 s,$$

for some strictly positive constants, $C_i, i = 1, ..., 6$, depending only on α. Therefore, for $t > 0$,

$$|\varepsilon_n(t)| \le |\varphi_\infty(t)| \int_0^t \frac{\left|\varphi_1^{n-1}\left(\frac{s}{n^{\frac{1}{\alpha}}}\right)\right|}{|\varphi_\infty(s)|} \left| n^{1-\frac{1}{\alpha}} \psi_1'\left(\frac{s}{n^{\frac{1}{\alpha}}}\right) + n^{1-\frac{1}{\alpha}} \psi_2'\left(\frac{s}{n^{\frac{1}{\alpha}}}\right) + n^{1-\frac{1}{\alpha}} \psi_3'\left(\frac{s}{n^{\frac{1}{\alpha}}}\right) - \alpha \frac{s^{2\alpha-1}}{n} \right.$$
$$\left. + \alpha s^{\alpha-1} \left(\psi_1\left(\frac{s}{n^{\frac{1}{\alpha}}}\right) + \psi_2\left(\frac{s}{n^{\frac{1}{\alpha}}}\right) + \psi_3\left(\frac{s}{n^{\frac{1}{\alpha}}}\right) \right) \right| ds$$
$$\le C|\varphi_\infty(t)| \int_0^t \frac{\left|\varphi_1^{n-1}\left(\frac{s}{n^{\frac{1}{\alpha}}}\right)\right|}{|\varphi_\infty(s)|} \left(\frac{s}{n^{\frac{2}{\alpha}-1}} + \frac{s^2}{n^{\frac{3}{\alpha}-1}} + \frac{s^{2\alpha-1}}{n} + \frac{s^{\alpha+1}}{n^{\frac{2}{\alpha}}} \right) ds,$$

and so

$$|\varepsilon_n(n^{\frac{1}{\alpha}}t)| \le Cn|\varphi_\infty(n^{\frac{1}{\alpha}}t)| \int_0^t \frac{|\varphi_1^{n-1}(u)|}{|\varphi_\infty(n^{\frac{1}{\alpha}}u)|} \left(u + u^2 + u^{2\alpha-1} + u^{\alpha+1} \right) du.$$

Let us now detail how to bound the ratio $|\varphi_1^{n-1}(u)|/|\varphi_\infty\left(n^{1/\alpha}u\right)|$. For $0 < u \le t$

$$\frac{|\varphi_1^{n-1}(u)|}{|\varphi_\infty\left(n^{\frac{1}{\alpha}}u\right)|} \le e^{nu^\alpha + (n-1)\ln\varphi_1(u)}$$
$$\le e^{nu^\alpha + (n-1)(-u^\alpha + \psi_1(u) + \psi_2(u) + \psi_3(u))}$$
$$\le e e^{n(\psi_1(u) + \psi_2(u) + \psi_3(u))}.$$

By (4.22), we can choose $\eta \in (0,1)$ such that $0 < C(\eta) = \max\limits_{u\in(0,\eta)} (|\psi_1(u)| + |\psi_2(u)| + |\psi_3(u)|)/u^\alpha < 1$, since $\alpha \in (1,2)$. Then, for $0 < u \le t \le \eta$,

$$\frac{|\varphi_1^{n-1}(u)|}{|\varphi_\infty\left(n^{\frac{1}{\alpha}}u\right)|} \le e e^{nC(\eta)u^\alpha},$$

which implies that, for $0 < t \le \eta < 1$,

$$|\varepsilon_n(n^{\frac{1}{\alpha}}t)| \le Cne^{-n(1-C(\eta))t^\alpha} \left(t^2 + t^3 + t^{2\alpha} + t^{\alpha+2} \right).$$

A similar bound can also be obtained for $-\eta \le t < 0$. Setting $T := n^{1/\alpha}\eta$, applying Esseen's inequality, and if h_α denotes the density of the $S\alpha S$-law, we finally get

$$d_K(X_n, X_\infty) \le C_1' \int_{-n^{1/\alpha}\eta}^{+n^{1/\alpha}\eta} \frac{|\varepsilon_n(t)|}{|t|} dt + C_2 \frac{\|h_\alpha\|_\infty}{n^{\frac{1}{\alpha}}\eta}$$
$$\le C_1' \int_{-\eta}^{+\eta} \frac{|\varepsilon_n(n^{\frac{1}{\alpha}}t)|}{|t|} dt + C_2 \frac{\|h_\alpha\|_\infty}{n^{\frac{1}{\alpha}}\eta}$$
$$\le C_1' \int_0^{+\eta} Cne^{-n(1-C(\eta))t^\alpha} \left(t + t^2 + t^{2\alpha-1} + t^{\alpha+1} \right) dt + C_2 \frac{\|h_\alpha\|_\infty}{n^{\frac{1}{\alpha}}\eta}$$

$$\leq C_1' \int_0^{n^{\frac{1}{\alpha}}\eta} e^{-(1-C(\eta))t^\alpha} \left(\frac{t}{n^{\frac{2}{\alpha}-1}} + \frac{t^2}{n^{\frac{3}{\alpha}-1}} + \frac{t^{2\alpha-1}}{n} + \frac{t^{\alpha+1}}{n^{\frac{2}{\alpha}}} \right) dt + C_2 \frac{\|h_\alpha\|_\infty}{n^{\frac{1}{\alpha}}\eta}$$

$$\leq C_{\eta,\alpha} \left(\frac{1}{n^{\frac{2}{\alpha}-1}} + \frac{1}{n^{\frac{3}{\alpha}-1}} + \frac{1}{n} + \frac{1}{n^{\frac{2}{\alpha}}} \right) + C_2 \frac{\|h_\alpha\|_\infty}{n^{\frac{1}{\alpha}}\eta}$$

$$\leq \frac{C_{\eta,\alpha,h_\alpha}}{n^{\frac{2}{\alpha}-1}},$$

for some $C_{\eta,\alpha,h_\alpha} > 0$ depending only on η, on α and $\|h_\alpha\|_\infty$. This concludes the proof of the proposition. □

Remark 4.8 *The above result has to be compared with the ones available in the literature but for other types of Pareto laws. Very recently, and via Stein's method, a rate of convergence in Wasserstein-1 and for symmetric stable limiting laws is obtained in [94]. When specialized to the Pareto law with density $f(x) := \alpha\lambda^\alpha/(2|x|^{\alpha+1})\mathbb{1}_{|x|>\lambda}$, described in the previous remark, this rate is of order $n^{-(2/\alpha-1)}$ (see [94]), which via the inequality (4.9) provides a rate of the order $n^{-(1/\alpha-1/2)}$ in Kolmogorov distance. Moreover, a convergence rate of order $n^{-(2/\alpha-1)}$ in Kolmogorov distance is known to hold for the same Pareto law (see, e.g., [33] and references therein). The results of [50] also imply a rate of convergence of order $n^{-(2/\alpha-1)}$ in Kolmogorov distance for the Pareto law considered in Proposition 4.7 (see [50, Corrolary 1]). At a different level, the rate $n^{-(2/\alpha-1)}$ also appears when one considers the convergence, in supremum norm, of the corresponding densities toward the stable density, see [60]. Finally, note also that Proposition 4.7 is a special case of [57, Theorem 1.2] proved with different techniques.*

Remark 4.9 *Analyzing the proof of Proposition 4.7, it is clearly possible to generalize the previous result beyond the Pareto case, to more general distributions pertaining to the domain of normal attraction of the symmetric α-stable distribution. Indeed, consider distribution functions of the form*

$$\forall x > 0, \quad F(x) = 1 - \frac{(c + a(x))}{x^\alpha},$$

$$\forall x < 0, \quad F(x) = \frac{(c + a(-x))}{(-x)^\alpha},$$

where the function a defined on $(0, +\infty)$ is such that $\lim_{x\to+\infty} a(x) = 0$, and where $c = (1-\alpha)/(2\Gamma(2-\alpha)\cos(\alpha\pi/2))$. Moreover, let a be bounded and continuous on $(0, +\infty)$ and be such that $\lim_{x\to+\infty} xa(x) < +\infty$. Then, by straightforward computations

$$\int_{-\infty}^{+\infty} e^{itx} dF(x) = 1 - t^\alpha + \psi(t),$$

where ψ is a real-valued function satisfying for all $t \in \mathbb{R}$, $|\psi(t)| \leq C_1|t|^2$ and $|\psi'(t)| \leq C_2|t|$, for some $C_1 > 0$ and $C_2 > 0$, two constants only depending *on α and a. Assuming further that the probability measure associated with the distribution function F is infinitely divisible, it follows that for $X_n = \sum_{i=1}^{n} \xi_i / n^{1/\alpha}$, with ξ_i iid random variables such that $\xi_1 \sim F$,*

$$d_K(X_n, X_\infty) \leq \frac{C}{n^{\frac{2}{\alpha}-1}},$$

where $X_\infty \sim S\alpha S$ and where $C > 0$ only depends on α and a.

A further simple adaptation of the proofs of the previous results leads to explicit rates of convergence for the compound Poisson approximation of some classes of infinitely divisible distributions. The next two results give Berry–Esseen-type bounds.

Theorem 4.10 *Let $X \sim ID(b, 0, \nu)$ be nondegenerate such that $\mathbb{E}|X| < \infty$ and such that*

$$\int_{-1}^{+1} |u| \nu(du) < \infty. \tag{4.23}$$

Let its law be absolutely continuous with respect to the Lebesgue measure with a bounded density, and let its characteristic function, φ, be such that, for all $t \in \mathbb{R}$,

$$|\varphi(t)| \int_0^{|t|} \frac{ds}{|\varphi(s)|} \leq C|t|^p, \tag{4.24}$$

where $C > 0$ and $p \geq 1$. Finally, let X_n, $n \geq 1$, be compound Poisson random variables each with characteristic function,

$$\varphi_n(t) := \exp\left(n\left((\varphi(t))^{\frac{1}{n}} - 1\right)\right). \tag{4.25}$$

Then,

$$d_K(X_n, X) \leq C' \left(\frac{1}{n}\right)^{\frac{1}{p+2}} \left(|b_0| + \int_{-\infty}^{+\infty} |u| \nu(du)\right)^{\frac{2}{p+2}}, \tag{4.26}$$

where $C' > 0$ depends on the supremum norm of the density of X, but is independent of n.

Proof Clearly, $\varphi_n(t) \to \varphi(t)$, for all $t \in \mathbb{R}$, and so the sequence $(X_n)_{n \geq 1}$ converges in distribution toward X. Then, adopting the notations of the proof of Theorem 4.1,

$$\frac{1}{i} \frac{d}{dt}(\varphi_n(t)) = \varphi_n(t)S(t) + \varphi_n(t)\Delta_n^+(t) - \varphi_n(t)\Delta_n^-(t).$$

Moreover, thanks to (4.25),

$$\frac{d}{dt}\big(\varphi_n(t)\big) = \frac{\varphi'(t)}{\varphi(t)}\big(\varphi(t)\big)^{\frac{1}{n}}\varphi_n(t) = iS(t)\big(\varphi(t)\big)^{\frac{1}{n}}\varphi_n(t),$$

since

$$S(t) = \frac{1}{i\varphi(t)}\frac{d}{dt}\big(\varphi(t)\big).$$

Thus,

$$\Delta_n^+(t) - \Delta_n^-(t) = S(t)\left(\big(\varphi(t)\big)^{\frac{1}{n}} - 1\right),$$

which implies that

$$\varepsilon_n(t) = i\varphi(t)\int_0^t \frac{\varphi_n(s)}{\varphi(s)}S(s)\left(\big(\varphi(s)\big)^{\frac{1}{n}} - 1\right)ds.$$

Therefore,

$$|\varepsilon_n(t)| \le |\varphi(t)|\int_0^t \frac{1}{|\varphi(s)|}|S(s)||\big(\varphi(s)\big)^{\frac{1}{n}} - 1|ds. \tag{4.27}$$

Next, by the very definition of S, Corollary 3.5, and straightforward computations,

$$S(s) = m_0^+\varphi_{Y^+}(s) - m_0^-\varphi_{Y^-}(s)$$
$$= \left(b_0^+ + \int_{-\infty}^{+\infty} e^{isu}\tilde{\nu}_+(du)\right) - \left(b_0^- + \int_{-\infty}^{+\infty} e^{isu}\tilde{\nu}_-(du)\right)$$
$$= b_0 + \int_{-\infty}^{+\infty} e^{isu}\tilde{\nu}(du).$$

Hence,

$$|S(s)| \le |b_0| + \int_{-\infty}^{+\infty} |u|\nu(du), \tag{4.28}$$

and further straightforward computations lead to

$$|\big(\varphi(s)\big)^{\frac{1}{n}} - 1| \le \frac{|s|}{n}\left(|b_0| + \int_{-\infty}^{+\infty} |u|\nu(du)\right). \tag{4.29}$$

Combining (4.24) and (4.27)–(4.29) gives

$$|\varepsilon_n(t)| \le C\frac{1}{n}\left(|b_0| + \int_{-\infty}^{+\infty} |u|\nu(du)\right)^2 |t|^{p+1}. \tag{4.30}$$

To conclude the proof of this theorem, proceed as in the end of the proof of Theorem 4.1. □

Proposition 4.11 *Let $X \sim ID(b, 0, \nu)$ be nondegenerate and such that $\mathbb{E}|X|^2 < \infty$. Let its law be absolutely continuous with respect to the Lebesgue measure with a bounded density, and let its characteristic function, φ, be such that, for all $t \in \mathbb{R}$,*

$$|\varphi(t)| \int_0^{|t|} \frac{ds}{|\varphi(s)|} \le C|t|^p, \qquad (4.31)$$

where $C > 0$ and $p \ge 1$. Finally, let X_n, $n \ge 1$, be compound Poisson random variables each with characteristic function,

$$\varphi_n(t) := \exp\left(n\left((\varphi(t))^{\frac{1}{n}} - 1\right)\right). \qquad (4.32)$$

Then,

$$d_K(X_n, X) \le C'\left(\frac{1}{n}\right)^{\frac{1}{p+4}}\left(|\mathbb{E}X| + \int_{-\infty}^{+\infty} u^2 \nu(du)\right)^{\frac{2}{p+4}}, \qquad (4.33)$$

where $C' > 0$ is independent of n.

Proof The proof is similar to the proof of Theorem 4.10 and so is only sketched. Clearly, $\varphi_n(t) \to \varphi(t)$, for all $t \in \mathbb{R}$ and so the sequence $(X_n)_{n \ge 1}$ converges in distribution toward X. Then, with the previous notations,

$$\frac{d}{dt}(\varphi_n(t)) = R(t)\varphi_n(t) + \Delta_n(t)\varphi_n(t).$$

Moreover, thanks to (4.32),

$$\frac{d}{dt}(\varphi_n(t)) = \frac{\varphi'(t)}{\varphi(t)}(\varphi(t))^{\frac{1}{n}}\varphi_n(t) = R(t)(\varphi(t))^{\frac{1}{n}}\varphi_n(t),$$

since

$$R(t) = \frac{1}{\varphi(t)}\frac{d}{dt}(\varphi(t)).$$

Thus,

$$\Delta_n(t) = R(t)\left((\varphi(t))^{\frac{1}{n}} - 1\right),$$

which implies that

$$\varepsilon_n(t) = \varphi(t) \int_0^t \frac{\varphi_n(s)}{\varphi(s)} R(s) \left((\varphi(s))^{\frac{1}{n}} - 1 \right) ds,$$

and therefore,

$$|\varepsilon_n(t)| \le |\varphi(t)| \int_0^t \frac{1}{|\varphi(s)|} |R(s)| |(\varphi(s))^{\frac{1}{n}} - 1| ds. \tag{4.34}$$

By the very definition of R, Proposition 3.8, and straightforward computations,

$$R(s) = -s \left(\int_{-\infty}^{+\infty} u^2 \nu(du) \right) \varphi_Y(s) + i\mathbb{E}X$$

$$= i \int_{-\infty}^{+\infty} \left(e^{isu} - 1 \right) u \nu(du) + i\mathbb{E}X.$$

Hence,

$$|R(s)| \le |\mathbb{E}X| + |s| \int_{-\infty}^{+\infty} u^2 \nu(du).$$

Moreover, further straightforward computations lead to

$$|(\varphi(s))^{\frac{1}{n}} - 1| \le \frac{|s|}{n} \left(|\mathbb{E}X| + |s| \int_{-\infty}^{+\infty} v^2 \nu(dv) \right). \tag{4.35}$$

Combining (4.31) and (4.34)–(4.35) gives

$$|\varepsilon_n(t)| \le C \frac{1}{n} \left(|\mathbb{E}X| + |t| \int_{-\infty}^{+\infty} u^2 \nu(du) \right)^2 |t|^{p+1},$$

$$\le C \frac{1}{n} \left(|\mathbb{E}X| + \int_{-\infty}^{+\infty} u^2 \nu(du) \right)^2 (|t|^{p+1} + |t|^{p+2} + |t|^{p+3}).$$

To conclude the proof of this proposition, proceed as in the end of the proof of Proposition 4.10. □

Remark 4.12 *(i) Under the condition*

$$A := \sup_{s \in \mathbb{R}} \left| \int_{-\infty}^{+\infty} (e^{isu} - 1) u \nu(du) \right| < \infty,$$

the upper bound on the Kolmogorov distance in Proposition 4.11 becomes

$$d_K(X_n, X) \leq C' \left(\frac{1}{n}\right)^{\frac{1}{p+2}} (|\mathbb{E}X| + A)^{\frac{2}{p+2}}, \tag{4.36}$$

which is comparable to the one obtained in Theorem 4.10 and is, for instance, verified in case the Lévy measure of X satisfies the assumptions of Theorem 4.10.

(ii) Once again, versions of Theorem 4.10 and of Proposition 4.11 can be derived for the smooth Wasserstein distance. Let us develop this claim a little bit more, and assume that X has finite exponential moments, namely, that $\mathbb{E}e^{\lambda|X|}$ is finite for some $\lambda > 0$. This condition implies that the characteristic function φ is analytic in a horizontal strip of the complex plane containing the real axis. Then, by the very definition of φ_n and the use of the Lévy–Raikov Theorem (see, e.g., [66, Theorem 10.1.1]), it follows that φ_n is analytic in at least the same horizontal strip. Moreover, in this strip, still by its very definition, $(\varphi_n)_{n\geq 1}$ converges pointwise toward φ. Hence, the random variables $e^{\eta|X_n|}$ (for some $\eta > 0$) are uniformly integrable. Therefore, if $\mathbb{E}e^{\lambda|X|}$ is finite and if the assumptions (4.23) and (4.24) hold true, (4.30) and [4, Theorem 1] lead to

$$d_{W_{p+2}}(X_n, X) \leq C \frac{\sqrt{\ln n}}{n} \left(|b_0| + \int_{-\infty}^{+\infty} |u|\nu(du)\right)^2, \tag{4.37}$$

for some constant C only depending on the limiting distribution.

We now present some examples illustrating the applicability of our methods, as developed to this point, by verifying the validity of various hypotheses.

(i) The gamma random variable with parameters $\alpha \geq 1$ and $\beta > 0$ satisfies the assumptions of Theorem 4.10. Indeed, (4.23) and the boundedness of the density are automatically verified, and moreover

$$|\varphi(t)| \int_0^{|t|} \frac{ds}{|\varphi(s)|} \leq |t|,$$

for all $t \in \mathbb{R}$.

(ii) Let $q \geq 3$ and let $(\lambda_1, ..., \lambda_q)$ be q nonzero distinct reals. Let $X := \sum_{k=1}^q \lambda_k(Z_k^2 - 1)$, where the $\{Z_i, i = 1, ..., q\}$ are iid standard normal random variables. Clearly, X is infinitely divisible and its Lévy measure is given by

$$\frac{\nu(du)}{du} := \left(\sum_{\lambda \in \Lambda_+} \frac{e^{-u/(2\lambda)}}{2u}\right) \mathbb{1}_{(0,+\infty)}(u) + \left(\sum_{\lambda \in \Lambda_-} \frac{e^{-u/(2\lambda)}}{2(-u)}\right) \mathbb{1}_{(-\infty,0)}(u),$$

where $\Lambda_+ = \{\lambda_k : \lambda_k > 0\}$ and $\Lambda_- = \{\lambda_k : \lambda_k < 0\}$ have finite cardinality, and so the condition (4.23) is verified. Moreover, $\varphi(t) := \prod_{j=1}^q e^{-it\lambda_j}/(1 - 2it\lambda_j)^{1/2}$ and thus

$$\frac{1}{\left(1+4\lambda_{\max}^2 t^2\right)^{\frac{q}{4}}} \le |\varphi(t)| \le \frac{1}{\left(1+4\lambda_{\min}^2 t^2\right)^{\frac{q}{4}}},$$

with $\lambda_{\max} = \max\limits_{k\ge 1} |\lambda_k|$ and $\lambda_{\min} = \min\limits_{k\ge 1} |\lambda_k|$. This readily implies that X has a bounded density and that, for all $t \in \mathbb{R}$,

$$|\varphi(t)| \int_0^{|t|} \frac{ds}{|\varphi(s)|} \le C|t|,$$

where

$$C := \sup_{t\in\mathbb{R}} \left(\frac{(1+4\lambda_{\max}^2 t^2)^{\frac{q}{4}}}{(1+4\lambda_{\min}^2 t^2)^{\frac{q}{4}}} \right) = \left(\frac{\lambda_{\max}}{\lambda_{\min}} \right)^{q/2}.$$

(iv) More generally, let $(\lambda_k)_{k\ge 1}$ be an absolutely summable sequence such that $|\lambda_k| \ne 0$, for all $k \ge 1$. Let $X := \sum_{k=1}^{+\infty} \lambda_k(Z_k^2 - 1)$ where $(Z_k)_{k\ge 1}$ is a sequence of iid standard normal random variables. Since $(\lambda_k)_{k\ge 1}$ is absolutely summable, the condition (4.23) is verified. Let us now fix $N \ge 3$ and assume that the absolute values of the eigenvalues $(\lambda_k)_{k\ge 1}$ are indexed in decreasing order, i.e., $|\lambda_1| \ge |\lambda_2| \ge \dots \ge |\lambda_N| \ge \dots$. Then, for all $t \in \mathbb{R}$,

$$\frac{\psi_N(t)}{\left(1+4t^2|\lambda_1|^2\right)^{\frac{N}{4}}} \le |\varphi(t)| \le \frac{\psi_N(t)}{\left(1+4t^2|\lambda_N|^2\right)^{\frac{N}{4}}},$$

where

$$\psi_N(t) := \prod_{k=N+1}^{+\infty} \frac{1}{(1+4t^2\lambda_k^2)^{\frac{1}{4}}}.$$

Since $0 \le \psi_N(t) \le 1$, it is clear that X has a bounded density. Moreover, for each N, ψ_N is a decreasing function, thus,

$$|\varphi(t)| \int_0^{|t|} \frac{ds}{|\varphi(s)|} \le C|t|, \tag{4.38}$$

with

$$C := \sup_{t\in\mathbb{R}} \left(\frac{(1+4\lambda_1^2 t^2)^{\frac{q}{4}}}{(1+4\lambda_N^2 t^2)^{\frac{q}{4}}} \right) = \left(\frac{\lambda_1}{\lambda_N} \right)^{q/2}.$$

The next theorem pertains to quantitative convergence results inside the second Wiener chaos. [73, Theorem 3.1] puts forward the fact that a sequence of second-order Wiener chaos random variables converging in law, necessarily converges toward a random variable which is the sum of a centered Gaussian random variable (possibly degenerate) and of an independent second-order Wiener chaos random variable. It is, therefore, natural to consider the following instances of convergence in law:

$$\sum_{k=1}^{+\infty} \lambda_{n,k}(Z_k^2 - 1) \underset{n \to +\infty}{\Longrightarrow} \sum_{k=1}^{+\infty} \lambda_{\infty,k}(Z_k^2 - 1).$$

To study this issue, below, ℓ^1 denotes the space of absolutely summable real-valued sequences and for any such sequence $\lambda = (\lambda_k)_{k \geq 1} \in \ell^1$, let $\|\lambda\|_{\ell^1} := \sum_{k=1}^{+\infty} |\lambda_k|$.

Theorem 4.13 *Let $(\lambda_n)_{n \geq 1}$ be a sequence of elements of ℓ^1, converging (in $\| \cdot \|_{\ell_1}$) toward $\lambda_\infty \in \ell^1$. Moreover, let $|\lambda_{\infty,k}| \neq 0$ and $|\lambda_{n,k}| \neq 0$, for all $k \geq 1$, $n \geq 1$, and further, let $\sum_{k=1}^{+\infty} \lambda_{n,k}^2 = \sum_{k=1}^{+\infty} \lambda_{\infty,k}^2 = 1/2$. Next, set $X_n = \sum_{k=1}^{+\infty} \lambda_{n,k}(Z_k^2 - 1)$, $n \geq 1$, $X_\infty = \sum_{k=1}^{+\infty} \lambda_{\infty,k}(Z_k^2 - 1)$, and let*

$$\Delta_n := |\|\lambda_n^+\|_{\ell^1} - \|\lambda_\infty^+\|_{\ell^1}| + |\|\lambda_n^-\|_{\ell^1} - \|\lambda_\infty^-\|_{\ell^1}|$$
$$+ \|\lambda_n^+ - \lambda_\infty^+\|_{\ell^1} + \|\lambda_n^- - \lambda_\infty^-\|_{\ell^1},$$

where $\Lambda_n^\pm := \{\lambda_{n,k}^\pm, k \geq 1\} = \{\lambda_{n,k}, \lambda_{n,k} > 0 (< 0)\}$, and similarly for $\Lambda_\infty^\pm$. Then,

$$d_K(X_n, X_\infty) \leq C'_\infty \sqrt{\Delta_n}. \tag{4.39}$$

and

$$d_{W_2}(X_n, X_\infty) \leq C''_\infty \Delta_n \sqrt{|\ln \Delta_n|}, \tag{4.40}$$

for some positive constants C'_∞, C''_∞ depending only on X_∞.

Proof Since, for each $n \geq 1$, λ_n is absolutely summable and since so is λ_∞, the conditions (4.1) of Theorem 4.1 are satisfied. Then, from the proof of Theorem 4.1 and, for all $t \geq 0$,

$$\varepsilon_n(t) = i\varphi_\infty(t) \int_0^t \frac{\varphi_n(s)}{\varphi_\infty(s)} \big(\Delta_n^+(s) - \Delta_n^-(s)\big) ds. \tag{4.41}$$

Let us compute the quantities Δ_n^+ and Δ_n^-. By definition (see Corollary 3.5),

$$b_0^n = -\int_{-\infty}^{+\infty} u\nu_n(du) = -\sum_{\lambda \in \Lambda_n^+} \lambda + \sum_{\lambda \in \Lambda_n^-} (-\lambda),$$

and

$$\tilde{\nu}_n^+(du) = \frac{1}{2}\sum_{\lambda\in\Lambda_n^+} e^{-\frac{u}{2\lambda}}\mathbb{1}_{(0,+\infty)}(u)du, \quad \tilde{\nu}_n^-(du) = \frac{1}{2}\sum_{\lambda\in\Lambda_n^-} e^{-\frac{u}{2\lambda}}\mathbb{1}_{(-\infty,0)}(u)du,$$

implying that

$$(m_0^n)^+ = (m_0^n)^- = \|\lambda_n\|_{\ell^1},$$

$$\varphi_{Y_n^+}(t) = \frac{1}{(m_0^n)^+}\left((b_0^n)^+ + \sum_{\lambda\in\Lambda_n^+}\frac{\lambda}{1-2it\lambda}\right),$$

$$\varphi_{Y_n^-}(t) = \frac{1}{(m_0^n)^-}\left((b_0^n)^- + \sum_{\lambda\in\Lambda_n^-}\frac{-\lambda}{1-2it\lambda}\right).$$

Then, after some straightforward computations,

$$\Delta_n^+(s) - \Delta_n^-(s) = b_0^n - b_0^\infty + \sum_{\lambda\in\Lambda_n^+}\frac{\lambda}{1-2is\lambda} - \sum_{\lambda\in\Lambda_\infty^+}\frac{\lambda}{1-2is\lambda}$$

$$+ \sum_{\lambda\in\Lambda_\infty^-}\frac{-\lambda}{1-2is\lambda} - \sum_{\lambda\in\Lambda_n^-}\frac{-\lambda}{1-2is\lambda}.$$

Therefore,

$$|\Delta_n^+(s) - \Delta_n^-(s)| \leq |\|\lambda_n^+\|_{\ell^1} - \|\lambda_\infty^+\|_{\ell^1}| + |\|\lambda_n^-\|_{\ell^1} - \|\lambda_\infty^-\|_{\ell^1}|$$

$$+ \|\lambda_n^+ - \lambda_\infty^+\|_{\ell^1} + \|\lambda_n^- - \lambda_\infty^-\|_{\ell^1}.$$

Combining the previous bound with (4.38) and (4.41) entails

$$|\varepsilon_n(t)| \leq C_\infty |t|\Delta_n.$$

Finally, proceeding as in the proof of Theorem 4.1 gives

$$d_K(X_n, X_\infty) \leq C'_\infty\sqrt{\Delta_n}.$$

As for the upper bound on the smooth Wasserstein distance, recall the following tail property of the second Wiener chaoses: there exists $K > 0$ such that for all unit variance X in the second Wiener chaos, and for all $x > 2$:

$$\mathbb{P}(|X| > x) \leq \exp(-Kx),$$

(see, e.g., [56, Theorem 6.7]). This tail estimate implies that

$$\sup_{n\geq 1} \mathbb{E}\, e^{\eta |X_n|} < \infty, \qquad \mathbb{E}\, e^{\eta_\infty |X_\infty|} < \infty,$$

for some $\eta, \eta_\infty > 0$, and [4, Theorem 1] finishes the proof of the theorem. $\qquad\square$

Remark 4.14 *In the previous theorem, one could also consider $(\lambda_n)_{n\geq 1}$ such that, for each $n \geq 1$, there exists $k_n \geq 1$ (converging to $+\infty$ with n) such that for all $k = 1, ..., k_n$, $|\lambda_{n,k}| \neq 0$ and for all $k \geq k_n + 1$, $\lambda_{n,k} = 0$. The quantity Δ_n would then depend on the remainder term, $R_n := \sum_{k=k_n+1}^{+\infty} |\lambda_{\infty,k}|$.*

To conclude this chapter on the compound Poisson approximation of infinitely divisible distributions let us consider the stable case. Clearly, and as already indicated, an α-stable random variable satisfies neither the hypotheses of Corollary 3.5 nor those of Proposition 3.8. Nevertheless, the identities (3.18) and (3.21) lead to our next result.

Theorem 4.15 *Let $\alpha \in (1, 2)$ and let X be an α–stable random variable with Lévy measure given by (2.10) where $c_1, c_2 \geq 0$ are such that $c_1 + c_2 > 0$, and with characteristic function φ. For each $n \geq 1$, let X_n be a compound Poisson random variables with characteristic function*

$$\varphi_n(t) := \exp\left(n\left((\varphi(t))^{\frac{1}{n}} - 1 \right) \right). \tag{4.42}$$

Then,

$$d_K(X_n, X) \leq C \frac{1}{n^{\frac{1}{1+\alpha}}}, \tag{4.43}$$

where $C > 0$ depends only on α, c_1 and c_2.

Proof Thanks to (3.21) with $f(\cdot) = e^{it\cdot}$,

$$\frac{d}{dt}(\varphi(t)) = i\varphi(t)\left(c_2 \int_0^{+\infty} (1 - e^{-itu})\frac{du}{u^\alpha} - c_1 \int_0^{+\infty} (1 - e^{itu})\frac{du}{u^\alpha} + \frac{c_1 - c_2}{\alpha - 1} \right). \tag{4.44}$$

Next, setting

$$S(t) := \left(c_2 \int_0^{+\infty} (1 - e^{-itu})\frac{du}{u^\alpha} - c_1 \int_0^{+\infty} (1 - e^{itu})\frac{du}{u^\alpha} + \frac{c_1 - c_2}{\alpha - 1} \right),$$

(4.42) gives

$$\frac{d}{dt}(\varphi_n(t)) = \frac{\varphi'(t)}{\varphi(t)}(\varphi(t))^{\frac{1}{n}}\varphi_n(t) = i S(t)(\varphi(t))^{\frac{1}{n}}\varphi_n(t).$$

Introducing the quantity $\Delta_n(t) := S(t)((\varphi(t))^{\frac{1}{n}} - 1)$,

$$\frac{d}{dt}(\varphi_n(t)) = iS(t)\varphi_n(t) + i\varphi_n(t)\Delta_n(t). \tag{4.45}$$

Subtracting (4.44) from (4.45) and setting $\varepsilon_n(t) = \varphi_n(t) - \varphi(t)$ lead to

$$\frac{d}{dt}(\varepsilon_n(t)) = iS(t)\varepsilon_n(t) + i\varphi_n(t)\Delta_n(t).$$

Thus, for $t \geq 0$,

$$\varepsilon_n(t) = i\varphi(t) \int_0^t \frac{\varphi_n(s)}{\varphi(s)} \Delta_n(s)ds,$$

and similarly for $t \leq 0$. Let us now bound S. For $s > 0$,

$$|S(s)| \leq c_2 \int_0^{+\infty} |1 - e^{-isu}| \frac{du}{u^\alpha} + c_1 \int_0^{+\infty} |1 - e^{isu}| \frac{du}{u^\alpha} + \frac{|c_1 - c_2|}{\alpha - 1}$$

$$\leq (c_1 + c_2)|s|^{\alpha-1} \frac{2^{2-\alpha}}{(2 - \alpha)(\alpha - 1)} + \frac{|c_1 - c_2|}{\alpha - 1},$$

using

$$\int_0^{+\infty} |1 - e^{isu}| \frac{du}{u^\alpha} \leq |s|^{\alpha-1} \left(\int_0^2 \frac{du}{u^{\alpha-1}} + 2 \int_2^{+\infty} \frac{du}{u^\alpha} \right)$$

$$\leq |s|^{\alpha-1} \frac{2^{2-\alpha}}{(2 - \alpha)(\alpha - 1)}.$$

Moreover,

$$\left| (\varphi(s))^{\frac{1}{n}} - 1 \right| \leq \frac{|s|}{n} \left((c_1 + c_2)|s|^{\alpha-1} \frac{2^{2-\alpha}}{(2 - \alpha)(\alpha - 1)} + \frac{|c_1 - c_2|}{\alpha - 1} \right),$$

and

$$\varphi(s) := \exp\left(is\mathbb{E}X - c|s|^\alpha \left(1 - i\beta \tan \frac{\pi\alpha}{2} \operatorname{sgn}(s) \right) \right),$$

with $c = c_1 + c_2$ and $\beta = (c_1 - c_2)/(c_1 + c_2)$. This implies that, for $t \geq 0$,

$$
\begin{aligned}
|\varepsilon_n(t)| &\le \frac{1}{n}|\varphi(t)| \int_0^t \frac{1}{|\varphi(s)|} \left((c_1 + c_2)s^{\alpha-1}\frac{2^{2-\alpha}}{(2-\alpha)(\alpha-1)} + \frac{|c_1 - c_2|}{\alpha-1} \right)^2 |s|\,ds \\
&\le \frac{1}{n}e^{-ct^\alpha} \int_0^t e^{cs^\alpha} s \left((c_1 + c_2)s^{\alpha-1}\frac{2^{2-\alpha}}{(2-\alpha)(\alpha-1)} + \frac{|c_1 - c_2|}{\alpha-1} \right)^2 ds \\
&\le \frac{2}{n}\left((c_1 + c_2)\frac{2^{2-\alpha}}{(2-\alpha)(\alpha-1)} + \frac{|c_1 - c_2|}{\alpha-1} \right)^2 (t + t^{2\alpha-1})\left(e^{-ct^\alpha} \int_0^t e^{cs^\alpha} ds \right) \\
&\le \frac{2C}{n}\left((c_1 + c_2)\frac{2^{2-\alpha}}{(2-\alpha)(\alpha-1)} + \frac{|c_1 - c_2|}{\alpha-1} \right)^2 \frac{(t^2 + t^{2\alpha})}{1 + t^\alpha},
\end{aligned}
$$

where (4.8) is used to obtain the last inequality and where $C > 0$ only depends on α, c_1 and c_2. Therefore, with a similar argument for $t \le 0$, it follows that, for all $t \in \mathbb{R}$,

$$
|\varepsilon_n(t)| \le \frac{C'(t^2 + |t|^{2\alpha})}{n\ 1 + |t|^\alpha},
$$

for some $C' > 0$ depending only on α and c. To conclude the proof of this theorem, we proceed as in the end of the proof of Theorem 4.10: by Esseen inequality,

$$
d_K(X_n, X) \le C_1 \int_{-T}^T \frac{|\varepsilon_n(t)|}{|t|}dt + C_2\frac{\|h_\alpha\|_\infty}{T},
$$

where $C_1 > 0$, $C_2 > 0$, while h_α is the density of the stable distribution. Thus,

$$
d_K(X_n, X) \le \frac{C_1'}{n}\left(\int_0^T \frac{t}{1+t^\alpha}dt + \int_0^T \frac{t^{2\alpha-1}}{1+t^\alpha}dt \right) + C_2\frac{\|h_\alpha\|_\infty}{T}.
$$

Next, the idea is to exploit the different behaviors of the functions $t \to t/(1 + t^\alpha)$ and $t \to t^{2\alpha-1}/(1 + t^\alpha)$ at 0 and at infinity to optimize in T appearing on the right-hand side of the previous inequality. Thus, for $T \ge 1$,

$$
\begin{aligned}
\int_0^T \frac{t}{1+t^\alpha}dt &\le \int_0^\varepsilon \frac{t}{1+t^\alpha}dt + \int_\varepsilon^T \frac{t}{1+t^\alpha}dt \\
&\le \frac{\varepsilon^2}{2} + \varepsilon^{1-\alpha}T \\
&\le CT^{\frac{2}{\alpha+1}},
\end{aligned}
$$

where we optimized in ε in the last line and where $C > 0$ only depends on α. Similarly,

$$
\int_0^T \frac{t^{2\alpha-1}}{1+t^\alpha}dt \le CT^\alpha.
$$

Thus,

$$d_K(X_n, X) \leq \frac{C}{n}\left(T^\alpha + T^{\frac{2}{\alpha+1}}\right) + C_2\frac{\|h_\alpha\|_\infty}{T}.$$

Finally, choosing $T = n^{\frac{1}{1+\alpha}}$ finishes the proof of the theorem. $\qquad\square$

For the symmetric α-stable distribution, and in view of the proof of Proposition 4.7, the rate of convergence obtained above can be improved to $n^{1/\alpha}$. This is the content of the next result.

Theorem 4.16 *Let $\alpha \in (1, 2)$ and let $X \sim S\alpha S$ with characteristic function $\varphi(t) = \exp(-|t|^\alpha)$, $t \in \mathbb{R}$. Let X_n, $n \geq 1$, be a compound Poisson random variable with characteristic function*

$$\varphi_n(t) := \exp\left(n\left((\varphi(t))^{\frac{1}{n}} - 1\right)\right).$$

Then,

$$d_K(X_n, X) \leq \frac{C}{n^{\frac{1}{\alpha}}}, \tag{4.46}$$

where $C > 0$ only depends on α.

Proof From the proof of Proposition 4.7 (with its notations), for $t \geq 0$,

$$\varepsilon_n(t) = i\varphi(t)\int_0^t \frac{\varphi_n(s)}{\varphi(s)}\Delta_n(s)ds.$$

Moreover, for all $0 \leq s \leq t$,

$$|\Delta_n(s)| \leq \frac{s^{2\alpha-1}}{n},$$

hence,

$$\left|\varepsilon_n(n^{\frac{1}{\alpha}}t)\right| \leq Cn\left|\varphi\left(n^{\frac{1}{\alpha}}t\right)\right|\int_0^t \left|\frac{\varphi_n(n^{\frac{1}{\alpha}}s)}{\varphi(n^{\frac{1}{\alpha}}s)}\right||s|^{2\alpha-1}ds.$$

We next detail how to bound the ratio $\left|\varphi_n\left(n^{1/\alpha}s\right)/\varphi\left(n^{1/\alpha}s\right)\right|$. For $0 \leq s \leq t \leq 1$,

$$\left|\frac{\varphi_n\left(n^{\frac{1}{\alpha}}s\right)}{\varphi\left(n^{\frac{1}{\alpha}}s\right)}\right| \leq \exp\left(n(\exp(-s^\alpha) - 1 + s^\alpha)\right).$$

Now, pick $\eta \in (0, 1)$ such that $0 < C(\eta) = \max\limits_{s \in (0,\eta)} (\exp(-s^\alpha) - 1 + s^\alpha)/s^\alpha < 1$. Then,

$$\left| \varepsilon_n \left(n^{\frac{1}{\alpha}} t \right) \right| \le C n e^{-n(1-C(\eta))t^\alpha} t^{2\alpha}.$$

A similar bound can also be obtained for $t \le 0$. Setting $T := n^{1/\alpha}\eta$, and applying Esseen's inequality, we finally get

$$
\begin{aligned}
d_K(X_n, X_\infty) &\le C_1' \int_{-n^{\frac{1}{\alpha}}\eta}^{+n^{\frac{1}{\alpha}}\eta} \frac{|\varepsilon_n(t)|}{|t|} dt + C_2 \frac{\|h_\alpha\|_\infty}{n^{\frac{1}{\alpha}}\eta} \\
&\le C_1' \int_{-\eta}^{+\eta} \frac{|\varepsilon_n(n^{\frac{1}{\alpha}}t)|}{|t|} dt + C_2 \frac{\|h_\alpha\|_\infty}{n^{\frac{1}{\alpha}}\eta} \\
&\le C_1' \int_0^{+\eta} C n e^{-n(1-C(\eta))t^\alpha} t^{2\alpha-1} dt + C_2 \frac{\|h_\alpha\|_\infty}{n^{\frac{1}{\alpha}}\eta} \\
&\le C_1' \int_0^{n^{\frac{1}{\alpha}}\eta} e^{-(1-C(\eta))t^\alpha} \frac{t^{2\alpha-1}}{n} dt + C_2 \frac{\|h_\alpha\|_\infty}{\eta n^{\frac{1}{\alpha}}} \\
&\le \frac{C_{\eta,\alpha}}{n} + C_2 \frac{\|h_\alpha\|_\infty}{n^{\frac{1}{\alpha}}\eta} \\
&\le \frac{C_{\eta,\alpha,h_\alpha}}{n^{\frac{1}{\alpha}}},
\end{aligned}
$$

for some $C_{\eta,\alpha,h_\alpha} > 0$, depending only on η, α and on $\|h_\alpha\|_\infty$, where, again, h_α is the (bounded) density of the $S\alpha S$-law. This concludes the proof of the theorem. $\quad\square$

Chapter 5
Solution to Stein's Equation
for Self-Decomposable Laws

Having found in Chap. 3 that the operator $\mathcal{A}_{\text{gen}}$ given for all $f \in BLip(\mathbb{R})$, by

$$\mathcal{A}_{\text{gen}} f(x) = xf(x) - bf(x) - \int_{-\infty}^{+\infty} (f(x+u) - f(x)\mathbb{1}_{|u|\leq 1})u\nu(du),$$

characterizes $X \sim ID(b, 0, \nu)$, the usual next step in Stein's method is now to show that for any $h \in \mathcal{H}$ (a class of nice functions), the equation

$$\mathcal{A}_{\text{gen}} f(x) = h(x) - \mathbb{E}h(X) \tag{5.1}$$

has a solution f_h which also belongs to a class of nice functions. Of course for $X \sim ID(b, \sigma^2, \nu)$, the integral operator $\mathcal{A}_{\text{gen}}$ becomes an integro-differential operator given by

$$\mathcal{A}_{\text{gen}} f(x) = xf(x) - \sigma^2 f'(x) - bf(x) - \int_{-\infty}^{+\infty} (f(x+u) - f(x)\mathbb{1}_{|u|\leq 1})u\nu(du)).$$

Then, when interested in comparing the law of some random variable Y with the law of X, one needs to estimate

$$\sup_{h\in\mathcal{H}} |\mathbb{E}h(Y) - \mathbb{E}h(X)| = \sup_{h\in\mathcal{H}} |\mathbb{E}\mathcal{A}_{\text{gen}} f_h(Y)|.$$

In the sequel, we develop a semigroup methodology to solve a corresponding Stein equation for nondegenerate self-decomposable laws on $\mathbb{R}$. Semigroup methods have been initiated in [9, 48] and mainly developed for multivariate normal approximation or for diffusions approximation. To start with, recall that, by definition, X, with characteristic function φ, is self-decomposable if for any $\gamma \in (0, 1)$,

© The Author(s), under exclusive license to Springer Nature Switzerland AG 2019
B. Arras and C. Houdré, *On Stein's Method for Infinitely Divisible Laws with Finite First Moment*, SpringerBriefs in Probability and Mathematical Statistics,
https://doi.org/10.1007/978-3-030-15017-4_5

$$\varphi_\gamma(t) := \frac{\varphi(t)}{\varphi(\gamma t)}, \tag{5.2}$$

$t \in \mathbb{R}$, is itself a characteristic function ([84, Definition 15.1]). Recall also that non-degenerate self-decomposable laws are infinitely divisible, absolutely continuous with respect to the Lebesgue measure (see [84, Proposition 15.5] and [87, Chap. V, Sect. 6, Theorem 6.14]) and moreover closed under convolution. The class of self-decomposable distributions comprises many of the infinitely divisible ones. To name but a few, the stable distributions, the gamma distributions, the second Wiener chaos-type distributions, the Laplace distribution, the generalized Dickman distribution, the double-Pareto distribution with $r > 1$, the log-normal distribution, the logistic distribution, the Student distribution with $r > 0$, the generalized inverse Gaussian, the generalized hyperbolic distributions (see, e.g., [49]), the half-Cauchy distribution (see, e.g., [36]), the Weibull distribution with parameter $0 < \alpha \le 1$ and the generalized gamma distribution with parameters (r, α) such that $r > 0$, $|\alpha| \le 1$ $\alpha \ne 0$ are all self-decomposable. We refer the reader to [84, 87] for more examples and properties of self-decomposable distributions.

To continue, and as usual, denote by $\mathcal{S}(\mathbb{R})$ the Schwartz space of infinitely differentiable rapidly decreasing real-valued functions defined on $\mathbb{R}$, and by $\mathcal{F}$ the Fourier transform operator given, for $f \in \mathcal{S}(\mathbb{R})$, by

$$\mathcal{F}(f)(\xi) = \int_{-\infty}^{+\infty} f(x)e^{-ix\xi}dx. \tag{5.3}$$

For $X \sim ID(b, 0, \nu)$, let also

$$\beta^* := \sup\left\{\beta \ge 1 : \int_{|u|>1} |u|^\beta d\nu(u) < +\infty\right\}. \tag{5.4}$$

For $X \sim ID(b, 0, \nu)$ nondegenerate and self-decomposable with law μ_X, let f_X be its Radon–Nikodym derivative with respect to the Lebesgue measure, let $S(f_X) = \{x \in \mathbb{R}, 0 \le f_X(x) < +\infty\}$ and let $N(f_X) = \{x \in \mathbb{R}, f_X(x) = 0\}$. Thanks to [84, Theorem 28.4], f_X is continuous on $S(f_X)$ (actually on $\mathbb{R}$ or on $\mathbb{R} \setminus \{b_0\}$, if b_0 exists). The integrability properties of the measure $|u|\nu(du)$ on $\{|u| \le 1\}$, ensure that the following alternatives hold true, e.g., see [84, Chap. 5, Sect. 24]:

- If $\int_{|u|\le 1} |u|\nu(du) < +\infty$, then the support of μ_X (denoted by $\text{Supp}(\mu_X)$) is either $[b_0, +\infty)$ or $(-\infty, b_0]$ or $\mathbb{R}$.
- If $\int_{|u|\le 1} |u|\nu(du) = +\infty$, then the support of μ_X is $\mathbb{R}$.

Before solving the Stein equation (5.1), let us start with the following proposition.

Proposition 5.1 *Let $X \sim ID(b, 0, \nu)$ be self-decomposable with law μ_X, characteristic function φ and such that $\mathbb{E}|X| < \infty$. Let $(P_t^\nu)_{t\ge 0}$ be the family of operators defined, for all $t \ge 0$ and for all $f \in \mathcal{S}(\mathbb{R})$, via*

$$P_t^\nu(f)(x) = \frac{1}{2\pi} \int_{-\infty}^{+\infty} \mathcal{F}(f)(\xi)e^{i\xi xe^{-t}} \frac{\varphi(\xi)}{\varphi(e^{-t}\xi)} d\xi. \tag{5.5}$$

Then, μ_X is invariant for $(P_t^\nu)_{t\geq 0}$, and $(P_t^\nu)_{t\geq 0}$ extends to a C_0-semigroup on $L^p(\mu_X)$, with $1 \leq p \leq \beta^*$. Its generator $\mathcal{A}$ is defined for all $f \in \mathcal{S}(\mathbb{R})$ and for all $x \in \mathbb{R}$ by

$$\mathcal{A}(f)(x) = \frac{1}{2\pi} \int_{-\infty}^{+\infty} \mathcal{F}(f)(\xi)e^{i\xi x}(i\xi)\left(-x + \mathbb{E}X + \int_{-\infty}^{+\infty} \left(e^{iu\xi} - 1\right)u\nu(du)\right)d\xi \tag{5.6}$$

$$= (\mathbb{E}X - x)f'(x) + \int_{-\infty}^{+\infty} \left(f'(x+u) - f'(x)\right)u\nu(du). \tag{5.7}$$

Proof First, it is easy to see that for any $f \in \mathcal{S}(\mathbb{R})$

$$P_0^\nu(f)(x) = f(x), \qquad \lim_{t\to+\infty} P_t^\nu(f)(x) = \int_{-\infty}^{+\infty} f(x)\mu_X(dx),$$

$$\int_{-\infty}^{+\infty} P_t^\nu(f)(x)\mu_X(dx) = \int_{-\infty}^{+\infty} f(x)\mu_X(dx).$$

Next, let $s, t \geq 0$ and $f \in \mathcal{S}(\mathbb{R})$. Then, on the one hand,

$$P_{t+s}^\nu(f)(x) = \frac{1}{2\pi} \int_{-\infty}^{+\infty} \mathcal{F}(f)(\xi)e^{i\xi e^{-(t+s)}x} \frac{\varphi(\xi)}{\varphi(e^{-(t+s)}\xi)} d\xi,$$

while, on the other hand,

$$\begin{aligned}
P_t^\nu(P_s^\nu(f))(x) &= \frac{1}{2\pi} \int_{-\infty}^{+\infty} \mathcal{F}(P_s^\nu(f))(\xi)e^{i\xi e^{-t}x} \frac{\varphi(\xi)}{\varphi(e^{-t}\xi)} d\xi \\
&= \frac{1}{2\pi} \int_{-\infty}^{+\infty} e^s \mathcal{F}(f)(e^s\xi) \frac{\varphi(e^s\xi)}{\varphi(\xi)} e^{i\xi e^{-t}x} \frac{\varphi(\xi)}{\varphi(e^{-t}\xi)} d\xi \\
&= \frac{1}{2\pi} \int_{-\infty}^{+\infty} \mathcal{F}(f)(\xi)e^{i\xi e^{-(t+s)}x} \frac{\varphi(\xi)}{\varphi(e^{-(t+s)}\xi)} d\xi,
\end{aligned}$$

since $\mathcal{F}(P_t^\nu(f))(\xi) = e^t \mathcal{F}(f)(e^t\xi)\dfrac{\varphi(e^t\xi)}{\varphi(\xi)}$. The semigroup property is therefore verified on $\mathcal{S}(\mathbb{R})$. Now, let $t \in (0, 1)$ and let $f \in \mathcal{S}(\mathbb{R})$. Then,

$$\frac{1}{t}\left(P_t^\nu(f)(x) - f(x)\right) = \int_{-\infty}^{+\infty} \mathcal{F}(f)(\xi)e^{i\xi x}\frac{1}{t}\left(e^{i\xi x(e^{-t}-1)}\frac{\varphi(\xi)}{\varphi(e^{-t}\xi)} - 1\right)\frac{d\xi}{2\pi}.$$

But, by Lemma A.1 of the Appendix,

$$\lim_{t \to 0^+} \frac{1}{t} \left(e^{i\xi x(e^{-t}-1)} \frac{\varphi(\xi)}{\varphi(e^{-t}\xi)} - 1 \right) = \left(-x + \mathbb{E}X + \int_{-\infty}^{+\infty} \left(e^{iu\xi} - 1 \right) u \nu(du) \right)(i\xi).$$

Moreover, applying Lemma A.2 of the Appendix, for $t \in (0, 1)$,

$$\left| \frac{1}{t} \left(e^{i\xi x(e^{-t}-1)} \frac{\varphi(\xi)}{\varphi(e^{-t}\xi)} - 1 \right) \right| \le C(1 + |\xi|)(\mathbb{E}|X| + |x| + |\xi| + 1),$$

for some constant $C > 0$, independent of t. Thus,

$$\lim_{t \to 0^+} \frac{1}{t} \left(P_t^\nu(f)(x) - f(x) \right) = \mathcal{A}(f)(x),$$

and, therefore, the generator of $(P_t^\nu)_{t \ge 0}$ on $\mathcal{S}(\mathbb{R})$ is indeed $\mathcal{A}$. Now, let $t \ge 0$, let $f \in \mathcal{S}(\mathbb{R})$ and let $1 \le p \le \beta^*$. Since X is self-decomposable, there exists, for each $t \ge 0$, a probability measure μ_t such that

$$\frac{\varphi(\xi)}{\varphi(e^{-t}\xi)} = \int_{-\infty}^{+\infty} e^{iu\xi} \mu_t(du), \qquad (5.8)$$

and thus,

$$P_t^\nu(f)(x) = \int_{-\infty}^{+\infty} f(u + e^{-t}x)\mu_t(du). \qquad (5.9)$$

The previous representation allows to extend the semigroup to $C_b(\mathbb{R})$, the space of bounded continuous functions on $\mathbb{R}$ endowed with the supremum norm, and $P_t^\nu(C_b(\mathbb{R})) \subset C_b(\mathbb{R})$. Therefore, P_t^ν is a contraction semigroup on $C_b(\mathbb{R})$ such that for all $f \in C_b(\mathbb{R})$ and all $t \ge 0$

$$\int_{\mathbb{R}} P_t^\nu(f)(x)\mu_X(dx) = \int_{\mathbb{R}} f(x)\mu_X(dx), \qquad (5.10)$$

and moreover such that, for all $f \in C_b(\mathbb{R})$, and all $x \in \mathbb{R}$,

$$\lim_{t \to 0^+} P_t^\nu(f)(x) = f(x). \qquad (5.11)$$

Indeed, one can check the invariance property (5.10) on $C_b(\mathbb{R})$ by noting that the probability measures $(\mu_X \otimes \mu_t) \circ \psi_t^{-1}$, with $\psi_t(x, y) = e^{-t}x + y$, and μ_X are the same. Similarly, one can check the pointwise convergence property (5.11) by using the fact that, by the Lévy continuity theorem, $\mu_t \circ \varphi_{t,x}^{-1}$, where $\varphi_{t,x}(u) = e^{-t}x + u$, for all $u \in \mathbb{R}$, converges weakly toward δ_x, as $t \to 0^+$. The semigroup property of $(P_t^\nu)_{t \ge 0}$ on $C_b(\mathbb{R})$ follows by similar arguments, and also,

$$\int_{-\infty}^{+\infty} \left| P_t^\nu(f)(x) \right|^p \mu_X(dx) \leq \int_{-\infty}^{+\infty} P_t^\nu(|f|^p)(x)\mu_X(dx)$$

$$\leq \int_{-\infty}^{+\infty} |f(x)|^p \mu_X(dx),$$

finishing the proof of our claim. Finally, a standard approximation argument concludes the proof of the proposition. □

Remark 5.2 *(i) It is important to note that the representation (5.9) also allows to extend the semigroup to the space of continuous functions on $\mathbb{R}$ vanishing at $\pm$ infinity. Moreover, this extension is a Feller semigroup (e.g., one can apply [80, Chap. III, Proposition 2.4]).*
(ii) Self-decomposable distributions are naturally associated with Ornstein–Uhlenbeck-type Markov processes. Indeed, thanks to [84, Theorem 17.5], for any self-decomposable distributions μ, one can find an Ornstein–Uhlenbeck-type Markov process such that μ is its invariant measure. Hence, from a heuristic point of view, it seems legitimate to implement a semigroup methodology to solve a Stein equation associated with a self-decomposable law.
(iii) There is a natural connection between generalized Mehler semigroups and self-decomposable probability measures. Indeed, for any probability measure on $\mathbb{R}$, let

$$D(\mu) := \{c \in [0, 1] : \mu = T_c(\mu) * \mu_c, \text{ for some } \mu_c \in M_1(\mathbb{R})\},$$

with $M_1(\mathbb{R})$ the set of probability measures on $\mathbb{R}$, with $T_c(\mu)(B) = \mu(B/c)$, for any Borel set B, and with the convention that $T_0(\mu) = \delta_0$. The set $D(\mu)$ is a closed multiplicative sub-semigroup containing 0 and 1. When μ is self-decomposable, this set is exactly $[0, 1]$; it contains in particular the one parameter semigroup $(e^{-t})_{t\geq 0}$. Note that $\mu_t = \mu_{e^{-t}}$, for all $t \geq 0$ with μ_t given by (5.8), which is well defined when μ is self-decomposable. The family of probability measures $(\mu_t)_{t\geq 0}$ then satisfies the following measure-valued cocycles (see [52]):

$$\mu_{s+t} = \mu_t * T_{e^{-t}}(\mu_s), \quad s, t > 0.$$

This convolution equality readily implies the semigroup property for the family of operators $(P_t^\nu)_{t\geq 0}$ of Lemma 5.1. For further information regarding the connection between decomposability of probability measures and generalized Mehler semigroups, the reader is referred to [58] and the references therein.
(iv) When ν is the Lévy measure of a symmetric α-stable distribution, the generator of the semigroup $(P_t^\nu)_{t\geq 0}$ boils down to

$$\mathcal{A}(f)(x) = -xf'(x) + \int_{-\infty}^{+\infty} \left(f'(x+u) - f'(x) \right) u\nu(du)$$

$$= -xf'(x) + c\int_0^{+\infty} \left(f'(x+u) - f'(x-u) \right) \frac{du}{u^\alpha},$$

which is, thanks to (3.20), proportional to the one considered in [94].
(v) For the generalized Dickman distribution, a Stein methodology has been developed in [17]. One of their Stein equations is (see [17, Equation (98)])

$$\frac{x}{\theta} f'_h(x) + f_h(x) - f_h(x+1) = h(x) - \mathbb{E}h(X), \qquad (5.12)$$

where X is a generalized Dickman random variable with parameter $\theta > 0$ and h belongs to the set of functions which are Lipschitz with Lipschitz derivative and with both Lipschitz constants at most one. Note that, by Proposition 5.1 and Example 3.1 (vii), the generator of the semigroup $(P^\nu_t)_{t\geq 0}$ boils down to

$$\mathcal{A}(f)(x) = (\theta - x)f'(x) + \int_{-\infty}^{+\infty} \left(f'(x+u) - f'(x) \right) u\nu(du)$$
$$= -xf'(x) + \theta(f(x+1) - f(x)),$$

which is clearly proportional to the differential-delay operator considered in [17, Equation (98)].
(vi) Let ν be the Lévy measure of the standard Laplace distribution, i.e., let $\nu(du) = |u|^{-1}e^{-|u|}du$, $u \neq 0$, then the generator of the associated semigroup boils down to

$$\mathcal{A}(f)(x) = -xf'(x) + \int_0^{+\infty} \left(f'(x+u) - f'(x-u) \right) e^{-u}du,$$

and the corresponding Stein's equation to

$$-xf'_h(x) + \int_0^{+\infty} \left(f'_h(x+u) - f'_h(x-u) \right) e^{-u}du = h(x) - \mathbb{E}h(X), \quad x \in \mathbb{R},$$
$$(5.13)$$

where X is the standard Laplace random variable. A Stein's method has also been developed for the Laplace distribution in [79]. There, the fundamental equation is

$$-f''_h(x) + f_h(x) = h(x) - \mathbb{E}h(X), \quad x \in \mathbb{R},$$

which does not seem to be immediately comparable to (5.13).
(vii) A Stein's method for exponential approximation is developed in [29, 42, 76]. There, the associated Stein equations read as

$$f'_h(x) - f_h(x) = h(x) - \mathbb{E}h(X), \quad x \geq 0,$$
$$xf'_h(x) - (x-1)f_h(x) = h(x) - \mathbb{E}h(X), \quad x > 0,$$

where now X is an exponential random variable with parameter 1, and h is an appropriate test function (not necessarily smooth). Since the Lévy measure of the *exponential distribution is* $\nu(du) = u^{-1}e^{-u}\mathbb{1}_{(0,+\infty)}(u)du$, *and thanks to Proposition 5.1, in our case the associated Stein equation is given, for all* $x \in \mathbb{R}$, *by*

$$(1-x)f_h'(x) + \int_0^{+\infty} \left(f_h'(x+u) - f_h'(x)\right)e^{-u}du = h(x) - \mathbb{E}h(X),$$

which is a nonlocal differential equation on the whole real line.
(viii) In [43, 65, 78], the following Stein equation has been introduced and analyzed, to study gamma approximation:

$$xf_h''(x) + (\alpha - \beta x)f_h'(x) = h(x) - \mathbb{E}h(X_{\alpha,\beta}), \quad x > 0,$$

where $X_{\alpha,\beta}$ *is a gamma random variable with parameters* $\alpha > 0, \beta > 0$ *and where* h *belongs to a suitable class of test functions. Also, in [37], the following version of the gamma Stein equation is used:*

$$xf_h'(x) + (\alpha - \beta x)f_h(x) = h(x) - \mathbb{E}h(X_{\alpha,\beta}), \quad x \in \mathbb{R},$$

but with the substantial difference that the above equation is now solved on the whole real line. In particular, in [37, Theorem 2.1], the following bounds on the solution are obtained when the test function h *is continuously differentiable on* $\mathbb{R}$ *and when both* h *and* h' *are Lipschitz:*

$$\|f_h\|_\infty \le \beta^{-1}\|h'\|_\infty, \quad \|f_h'\|_\infty \le 2\max\left(1, \frac{1}{\alpha}\right)\|h'\|_\infty, \quad \|f_h''\|_\infty \le 4\beta\max\left(1, \frac{1}{\alpha}\right)\|h'\|_\infty + 2\|h''\|_\infty.$$

Recalling that the Lévy measure of the gamma distribution is $\nu(du) = \alpha e^{-\beta u}u^{-1}\mathbb{1}_{(0,+\infty)}(u)du$, *the gamma Stein equation inferred from Proposition 5.1 is*

$$\left(\frac{\alpha}{\beta} - x\right)f_h'(x) + \int_0^{+\infty} \left(f_h'(x+u) - f_h'(x)\right)\alpha e^{-\beta u}du = h(x) - \mathbb{E}h(X_{\alpha,\beta}), \quad x \in \mathbb{R},$$

which is, once again, a nonlocal differential equation on the whole real line. Moreover, as shown below, when the test function h *is twice continuously differentiable on* $\mathbb{R}$ *and such that* $\|h'\|_\infty < +\infty$, $\|h''\|_\infty < +\infty$ *then, the following bounds, which are uniform in* $\alpha > 0$ *and* $\beta > 0$, *hold true:*

$$\|f_h'\|_\infty \le \|h'\|_\infty, \quad \|f_h''\|_\infty \le \frac{\|h''\|_\infty}{2}.$$

In particular, these bounds do not explode when $\alpha \to 0^+$.
(ix) Recall that since μ_X, *the law of* X, *is nondegenerate and self-decomposable, its Lévy measure admits the following representation* $\nu(du) = \psi(u)|u|^{-1}du$, $u \neq 0$, *where* ψ *is a nonnegative function increasing on* $(-\infty, 0)$ *and decreasing on* $(0, +\infty)$ *([84, Corollary 15.11]). Then, the probability measure* μ_t *(defined via (5.8)) is infinitely divisible. Denoting by* ν_t *the Lévy measure corresponding to* μ_t, *one easily*

checks that

$$\nu_t(du) = \frac{\psi(u) - \psi(e^t u)}{|u|} du.$$

With the help of the previous proposition, we now wish to solve the Stein equation associated with the operator $\mathcal{A}$. More precisely, for any Lipschitz function or bounded Lipschitz function h, we wish to solve the following integro-differential equation:

$$(\mathbb{E}X - x)f'(x) + \int_{-\infty}^{+\infty} \left(f'(x+u) - f'(x)\right) u\nu(du) = h(x) - \mathbb{E}h(X). \quad (5.14)$$

Using classical semigroup theory ([75, Chap. 2] or [41, Chap. 1]), the first step is to prove that

$$\int_0^{+\infty} \left(P_t^\nu(h)(x) - \mathbb{E}h(X)\right) dt$$

is well defined when h is a (bounded) Lipschitz function. At first, let h be a continuously differentiable function on $\mathbb{R}$ such that $\|h\|_\infty \leq 1$ and $\|h'\|_\infty \leq 1$. Since

$$|P_t^\nu(h)(x) - \mathbb{E}h(X)| = \left| \int_{-\infty}^{+\infty} h(y + e^{-t}x)\mu_t(dy) - \int_{-\infty}^{+\infty} h(y)\mu_X(dy) \right|$$

$$\leq \int_{-\infty}^{+\infty} |h(y + e^{-t}x) - h(y)|\mu_t(dy) + d_{W_1}(\mu_t, \mu_X)$$

$$\leq e^{-t}|x| + d_{W_1}(\mu_t, \mu_X),$$

we need to estimate the rate at which μ_t converges to μ_X in smooth Wasserstein-1 distance. We begin by estimating the rate at which μ_t converges toward μ_X in smooth Wasserstein-2 distance.

Proposition 5.3 *Let $X \sim ID(b, 0, \nu)$ be nondegenerate self-decomposable with law μ_X, and characteristic function φ, and moreover such that $\mathbb{E}|X| < \infty$. Let X_t, $t \geq 0$, be random variables each having characteristic function*

$$\varphi_t(\xi) = \frac{\varphi(\xi)}{\varphi(e^{-t}\xi)}, \qquad \xi \in \mathbb{R}. \quad (5.15)$$

Then,

$$d_{W_2}(X_t, X) \leq Ce^{-\frac{t}{4}}, \quad (5.16)$$

for $t > 0$ and for some $C > 0$ independent of t.

Proof One can apply [3, Theorem A.1] with $d = 1$ to get the exponential decay of $d_{W_2}(X_t, X)$. However, let us describe the proof in the univariate setting. Let X_t be a random variable with law μ_t given via (5.8). Then,

$$X_t =_d (1 - e^{-t})\mathbb{E}X + X_t^1 + X_t^2, \tag{5.17}$$

where X_t^1 and X_t^2 are independent and, respectively, defined, for all $\xi \in \mathbb{R}$ and for all $t > 0$, via

$$\mathbb{E}e^{i\xi X_t^1} = \exp\int_{|u|\le 1}\left(e^{iu\xi} - 1 - iu\xi\right)\nu_t(du), \quad \mathbb{E}e^{i\xi X_t^2} = \exp\int_{|u|>1}\left(e^{iu\xi} - 1 - iu\xi\right)\nu_t(du),$$

where ν_t is the Lévy measure of X_t. Moreover, from [67, inequality 13]

$$\mathbb{E}|X_t| \le (1 - e^{-t})\mathbb{E}|X| + \left(\int_{|u|\le 1}|u|^2\nu_t(du)\right)^{\frac{1}{2}} + 2\int_{|u|\ge 1}|u|\nu_t(du), \tag{5.18}$$

which implies, in particular, that $\sup_{t>0}\mathbb{E}|X_t| < +\infty$, since $\nu_t(B) \le \nu(B)$, for all Borel sets B. Let us next estimate the difference between φ_t and φ, the respective characteristic functions of μ_t and μ_X. For $t \ge 0$,

$$|\varphi_t(\xi) - \varphi(\xi)| \le \left|\frac{\varphi(\xi)}{\varphi(e^{-t}\xi)}\right||1 - \varphi(e^{-t}\xi)|$$

$$\le \mathbb{E}|X||\xi|e^{-t}.$$

Now, let g be an infinitely differentiable function with compact support contained in the interval $[-2R, 2R]$, for some $R > 1$. Then by Fourier inversion and Fubini theorem, for all $t > 0$,

$$|\mathbb{E}g(X) - \mathbb{E}g(X_t)| \le e^{-t}\mathbb{E}|X|\frac{1}{2\pi}\int_{\mathbb{R}}|\mathcal{F}(g)(\xi)||\xi|d\xi$$

$$\le e^{-t}\mathbb{E}|X|\frac{1}{2\pi}\int_{\mathbb{R}}|\mathcal{F}(g)(\xi)|\frac{(1 + |\xi|)^3}{(1 + |\xi|)^3}|\xi|d\xi$$

$$\le e^{-t}\mathbb{E}|X|\sup_{\xi\in\mathbb{R}}\left(|\mathcal{F}(g)(\xi)|(1 + |\xi|^3)\right)\left(\frac{1}{2\pi}\int_{\mathbb{R}}\frac{|\xi|d\xi}{(1 + |\xi|)^3}\right).$$

Moreover, for all $p \ge 2$

$$\sup_{\xi\in\mathbb{R}}\left(|\mathcal{F}(g)(\xi)|(1 + |\xi|^p)\right) \le CR\left(\|g\|_\infty + \|g^{(p)}\|_\infty\right),$$

for some $C > 0$. Thus, for all $t > 0$

$$|\mathbb{E}g(X) - \mathbb{E}g(X_t)| \le \tilde{C}e^{-t}\mathbb{E}|X|R\left(\|g\|_\infty + \|g^{(3)}\|_\infty\right), \tag{5.19}$$

with $\tilde{C} > 0$. Now, let $h \in C_c^\infty(\mathbb{R}) \cap \mathcal{H}_3$. Let Ψ_R be a compactly supported infinitely differentiable function on $\mathbb{R}$ whose support is contained in $[-2R, 2R]$, with values in $[0, 1]$ and such that $\Psi_R(x) = 1$, for all x such that $|x| \leq R$. Then, for all $t > 0$

$$|\mathbb{E}h(X) - \mathbb{E}h(X_t)| \leq |\mathbb{E}h(X)\Psi_R(X) - \mathbb{E}h(X_t)\Psi_R(X_t)| + |\mathbb{E}h(X)(1 - \Psi_R(X))|$$
$$+ |\mathbb{E}h(X_t)(1 - \Psi_R(X_t))|.$$

Now, note that

$$|\mathbb{E}h(X_t)(1 - \Psi_R(X_t))| \leq \int_{\mathbb{R}} (1 - \Psi_R(x)) d\mu_t(x)$$
$$\leq \mathbb{P}(|X_t| \geq R)$$
$$\leq \frac{1}{R} \sup_{t > 0} \mathbb{E}|X_t|,$$

which is finite by the first part of the proof. A similar bound holds true for $|\mathbb{E}h(X)(1 - \Psi_R(X))|$. Moreover, from (5.19),

$$|\mathbb{E}h(X) - \mathbb{E}h(X_t)| \leq \frac{C_1}{R} + \tilde{C}_1 e^{-t} \mathbb{E}|X| R \left(\|h\Psi_R\|_\infty + \|(h\Psi_R)^{(3)}\|_\infty \right),$$

for some constants $C_1, \tilde{C}_1 > 0$. Now,

$$\|h\Psi_R\|_\infty \leq 1,$$

and, by taking for Ψ_R an appropriate scaling of a bump function Ψ,

$$\|(h\Psi_R)^{(3)}\|_\infty \leq D,$$

for some $D > 0$ independent of R and h. Then,

$$|\mathbb{E}h(X) - \mathbb{E}h(X_t)| \leq C_2 \left(\frac{1}{R} + Re^{-t} \mathbb{E}|X| \right).$$

for some $C_2 > 0$. Choosing $R = e^{t/2}$, for all $t > 0$, it follows that

$$d_{W_3}(X, X_t) \leq \tilde{C}_2 e^{-\frac{t}{2}},$$

for some $\tilde{C}_2 > 0$. Using Lemma A.4 with $r = 3$,

$$d_{W_2}(X, X_t) \leq C_3 e^{-\frac{t}{4}},$$

for some $C_3 > 0$. This concludes the proof of the proposition. $\square$

Thanks to Lemma A.4 with $r = 2$, it is possible to link the smooth Wasserstein-2 distance to the smooth Wasserstein-1 distance, e.g.,

$$dw_1(X, Y) \leq 3\sqrt{2}\sqrt{dw_2(X, Y)},$$

for any two random variables X and Y. Therefore, combining Proposition 5.3 with Lemma A.4, with $r = 2$, yields

$$|P_t^\nu(h)(x) - \mathbb{E}h(X)| \leq e^{-t}|x| + Ce^{-\frac{t}{8}}, \tag{5.20}$$

which implies that $\int_0^{+\infty} |P_t^\nu(h)(x) - \mathbb{E}h(X)|dt < +\infty$, ensuring the well definiteness of the function

$$f_h(x) = -\int_0^{+\infty} \left(P_t^\nu(h)(x) - \mathbb{E}h(X)\right) dt, \quad x \in \mathbb{R}. \tag{5.21}$$

Let us now study the regularity of f_h.

Lemma 5.4 *Let h be a continuously differentiable function such that $\|h\|_\infty \leq 1$ and $\|h'\|_\infty \leq 1$. Then, f_h is differentiable on $\mathbb{R}$ and $\|f_h'\|_\infty \leq 1$.*

Proof Since

$$\frac{d}{dx}\left(P_t^\nu(h)(x)\right) = e^{-t}\int_{-\infty}^{+\infty} h'(xe^{-t} + y)\mu_t(dy),$$

it is clear that f_h is differentiable and that $\|f_h'\|_\infty \leq 1$. $\quad\square$

When h is a bounded Lipschitz function, the existence of higher order derivatives for f_h is linked to the behavior at $\pm\infty$ of the characteristic function of the probability measure μ_t as well as to heat kernel estimates for the density of μ_t. To illustrate these ideas, let us provide some examples.

(i) Let X be a gamma random variable with parameters $(\alpha, 1)$. Then, for all $\xi \in \mathbb{R}$,

$$\left|\frac{\varphi(\xi)}{\varphi(e^{-t}\xi)}\right| = \left(\frac{1 + e^{-2t}\xi^2}{1 + \xi^2}\right)^{\frac{\alpha}{2}},$$

is a decreasing function of ξ on $\mathbb{R}_+$ and thus,

$$e^{-\alpha t} \leq \left|\frac{\varphi(\xi)}{\varphi(e^{-t}\xi)}\right| \leq 1.$$

Similar upper and lower bounds hold for more general probability laws pertaining to the second Wiener chaos.

(ii) Let X be a Dickman random variable with parameter $\theta = 1$, namely, let the characteristic function of X be given, for all $\xi \in \mathbb{R}$, by

$$\varphi(\xi) = \exp\left(\int_0^1 \frac{e^{i\xi u} - 1}{u} du\right)$$

(see [4]). Then, for all $\xi \in \mathbb{R}$,

$$\frac{\varphi(\xi)}{\varphi(e^{-t}\xi)} = \exp\left(\int_0^1 e^{iu\xi} \frac{1 - e^{iu\xi(e^{-t}-1)}}{u} du\right).$$

Using standard asymptotic expansion for the cosine integral [74, Formulae 6.12.3, 6.12.4 and 6.2.20, Chap. 6], for all $\xi \in \mathbb{R}$,

$$C_1 \frac{1 + |\xi|e^{-t}}{1 + |\xi|} \le \left|\frac{\varphi(\xi)}{\varphi(e^{-t}\xi)}\right| \le C_2 \frac{1 + |\xi|e^{-t}}{1 + |\xi|},$$

for some $C_1 > 0, C_2 > 0$ independent of t.

(iii) Let X be a $S\alpha S$ random variable with $\alpha \in (1, 2)$. Then, for all $\xi \in \mathbb{R}$,

$$\frac{\varphi(\xi)}{\varphi(e^{-t}\xi)} = e^{-(1-e^{-\alpha t})|\xi|^\alpha}.$$

By Fourier inversion, μ_t admits a smooth density q such that, for all $k \ge 0$,

$$q^{(k)}(t, y) = \frac{1}{2\pi} \int_{-\infty}^{+\infty} \left(\frac{\varphi(\xi)}{\varphi(e^{-t}\xi)}\right) e^{iy\xi} (i\xi)^k d\xi.$$

Moreover, denoting by $p_\alpha(t, y)$ the probability transition density function of a one-dimensional α-stable Lévy process and noting that $q(t, y) = p_\alpha(1 - e^{-\alpha t}, y)$, Lemma 2.2 of [32] implies that

$$|q'(t, y)| \le C \frac{1 - e^{-\alpha t}}{\left((1 - e^{-\alpha t})^{\frac{1}{\alpha}} + |y|\right)^{\alpha+2}}.$$

As in the proof of Proposition 4.3 in [94], this last inequality implies that the function f_h admits a second-order derivative uniformly bounded such that

$$\|f_h''\|_\infty \le C_\alpha \|h'\|_\infty,$$

for some constant $C_\alpha > 0$, only depending on α.

The previous examples point out that for some specific target laws the semigroup solution to the Stein equation (5.14) with h bounded Lipschitz, might not reach second-order differentiability. Nevertheless, if h is a $C^2(\mathbb{R})$-function such that

$\|h\|_\infty$, $\|h'\|_\infty$, $\|h''\|_\infty \leq 1$, then f_h is twice differentiable with

$$\|f_h''\|_\infty \leq \frac{1}{2}.$$

In this instance, it is possible to partially transform the nonlocal part of the Stein operator into an operator acting on second derivatives. This is the purpose of the next lemma whose proof is very similar to the one of [94, Lemma 4.6] and, as such, is only sketched.

Lemma 5.5 *Let ν be a Lévy measure such that $\int_{|u|>1} |u|\nu(du) < +\infty$. Let f be a twice continuously differentiable function with first and second derivatives bounded. Then, for all $N > 0$, and all $x \in \mathbb{R}$,*

$$\int_{-\infty}^{+\infty} (f'(x+u) - f'(x))u\nu(du) = \int_{-N}^{+N} K_\nu(t, N) f''(x+t)dt + R_N(x),$$
(5.22)

where $K_\nu(t, N)$ and $R_N(x)$ are, respectively, given by

$$K_\nu(t, N) = \mathbb{1}_{[0,N]}(t) \int_t^N u\nu(du) + \mathbb{1}_{[-N,0]}(t) \int_{-N}^t (-u)\nu(du),$$

$$R_N(x) = \int_{|u|>N} (f'(x+u) - f'(x))u\nu(du).$$

Proof For $N > 0$ and $x \in \mathbb{R}$,

$$\int_{-\infty}^{+\infty} (f'(x+u) - f'(x))u\nu(du) = \int_0^N (f'(x+u) - f'(x))u\nu(du)$$

$$+ \int_{-N}^0 (f'(x+u) - f'(x))u\nu(du) + R_N(x).$$
(5.23)

For the first term on the right-hand side of (5.23),

$$\int_0^N (f'(x+u) - f'(x))u\nu(du) = \int_0^N \left(\int_0^u f''(x+t)dt\right) u\nu(du)$$

$$= \int_0^N f''(x+t)\left(\int_t^N u\nu(du)\right) dt, \quad (5.24)$$

while similar computations for the second term lead to

$$\int_{-N}^0 (f'(x+u) - f'(x))u\nu(du) = \int_{-N}^0 f''(x+t)\left(\int_{-N}^t (-u)\nu(du)\right) dt. \quad (5.25)$$

Combining (5.24) and (5.25) gives

$$\int_{-\infty}^{+\infty} (f'(x+u) - f'(x))u\nu(du) = \int_{-N}^{+N} K_\nu(t, N)f''(x+t)dt + R_N(x).$$

$\square$

Next, we study the regularity properties of the nonlocal part of the Stein operator. For this purpose, set

$$\mathcal{T}(f)(x) := \int_{-\infty}^{+\infty} \left(f'(x+u) - f'(x)\right)u\nu(du). \qquad (5.26)$$

Proposition 5.6 *Let ν be a Lévy measure such that $\int_{|u|>1} |u|\nu(du) < \infty$. Let f be a twice continuously differentiable function such that $\|f'\|_\infty < +\infty$ and $\|f''\|_\infty < +\infty$.*
(i) If $\int_{|u|\leq 1} |u|\nu(du) < +\infty$, then

$$\|\mathcal{T}(f)\|_{Lip} < +\infty.$$

(ii) If $\int_{|u|\leq 1} |u|\nu(du) = +\infty$ and if there exist $\gamma, \beta > 0$ and $C_1, C_2 > 0$ such that for any $R > 0$,

$$\int_{|u|>R} |u|\nu(du) \leq \frac{C_1}{R^\gamma}, \qquad \int_{|u|\leq R} |u|^2\nu(du) \leq C_2 R^\beta,$$

then,

$$\sup_{x\neq y} \frac{|\mathcal{T}(f)(x) - \mathcal{T}(f)(y)|}{|x-y|^{\frac{\beta}{\beta+\gamma}}} < +\infty.$$

Proof Let us start with (i). Let $x, y \in \mathbb{R}$, $x \neq y$. Then,

$$|\mathcal{T}(f)(x) - \mathcal{T}(f)(y)| \leq \left| \int_{-\infty}^{+\infty} \left(f'(x+u) - f'(y+u) - f'(x) + f'(y)\right) u\nu(du) \right|$$

$$\leq 2\|f''\|_\infty |x-y| \int_{-\infty}^{+\infty} |u|\nu(du),$$

showing that

$$\|\mathcal{T}(f)\|_{Lip} \leq 2 \int_{-\infty}^{+\infty} |u|\nu(du)\|f''\|_\infty. \qquad (5.27)$$

Let us next prove (ii). Let $R > 0$ and let $x, y \in \mathbb{R}$, $x \neq y$. Then,

$$|T(f)(x) - T(f)(y)| \leq \left| \int_{-\infty}^{+\infty} \left(f'(x+u) - f'(y+u) - f'(x) + f'(y) \right) u\nu(du) \right|$$

$$\leq \int_{|u| \leq R} \left| f'(x+u) - f'(y+u) - f'(x) + f'(y) \right| |u|\nu(du)$$

$$+ \int_{|u| > R} \left| f'(x+u) - f'(y+u) - f'(x) + f'(y) \right| |u|\nu(du).$$

$$\leq 2\|f''\|_\infty \left(\int_{|u| \leq R} |u|^2 \nu(du) + |x-y| \int_{|u| > R} |u|\nu(du) \right)$$

$$\leq 2C\|f''\|_\infty \left(R^\beta + \frac{|x-y|}{R^\gamma} \right).$$

Choosing $R = |x - y|^{\frac{1}{\gamma+\beta}}$ entails,

$$|T(f)(x) - T(f)(y)| \leq 2C\|f''\|_\infty |x-y|^{\frac{\beta}{\beta+\gamma}},$$

concluding the proof of the proposition. $\qquad\square$

Remark 5.7 *Above, when ν is the Lévy measure of a symmetric α-stable probability distribution, the exponents γ and β are, respectively, equal to $\alpha - 1$ and $2 - \alpha$ and so $\beta/(\gamma + \beta) = 2 - \alpha$ which is exactly the right order of Hölderian regularity needed for optimal approximation results as in [94].*

To end this chapter, we solve the Stein equation (5.14) for h infinitely differentiable with compact support such that $\|h\|_\infty \leq 1$, $\|h'\|_\infty \leq 1$ and $\|h''\|_\infty \leq 1$. This solution ensures, as shown in the results of the next chapter, via the representation (2.14), the existence of quantitative bounds on the smooth Wasserstein-2 distance. Note also that, from the Fourier approach in defining the semigroup $(P_t^\nu)_{t \geq 0}$, the function f_h is a solution to the Stein equation (5.14) on the whole real line.

Lemma 5.8 *Let $X \sim ID(b, 0, \nu)$ be nondegenerate, self-decomposable and such that $\mathbb{E}|X| < \infty$. Let $h \in C_c^\infty(\mathbb{R})$ be such that $\|h\|_\infty \leq 1$, $\|h'\|_\infty \leq 1$ and $\|h''\|_\infty \leq 1$. Let f_h be the function given by (5.21). Then, for all $x \in \mathbb{R}$,*

$$(\mathbb{E}X - x)f_h'(x) + \int_{-\infty}^{+\infty} \left(f_h'(x+u) - f_h'(x) \right) u\nu(du) = h(x) - \mathbb{E}h(X). \quad (5.28)$$

Proof Let $h \in C_c^\infty(\mathbb{R})$ be such that $\|h\|_\infty \leq 1$, $\|h'\|_\infty \leq 1$ and $\|h''\|_\infty \leq 1$ and let f_h be given by (5.21). Let $\hat{h} := h - \mathbb{E}h(X)$ and let $\psi \in \mathcal{S}(\mathbb{R})$. Now, by Fourier arguments as in the proof of Proposition 5.1

$$\frac{d}{dt}\langle P_t^\nu(h); \psi \rangle = \langle \mathcal{A}(P_t^\nu(h)); \psi \rangle, \quad (5.29)$$

where $\langle f; g \rangle = \int_{-\infty}^{+\infty} f(x)g(x)dx$. Thus, integrating from 0 to $+\infty$,

$$\int_0^{+\infty} \frac{d}{dt} \langle P_t^\nu(h); \psi \rangle dt = \int_0^{+\infty} \langle \mathcal{A}(P_t^\nu(h)); \psi \rangle dt. \tag{5.30}$$

Let us first deal with the left-hand side of (5.30). By definition,

$$\int_0^{+\infty} \frac{d}{dt} \langle P_t^\nu(h); \psi \rangle dt = \lim_{t \to +\infty} \langle P_t^\nu(h); \psi \rangle - \lim_{t \to 0^+} \langle P_t^\nu(h); \psi \rangle.$$

Straightforward applications of the dominated convergence theorem therefore imply that

$$\int_0^{+\infty} \frac{d}{dt} \langle P_t^\nu(h); \psi \rangle dt = \langle \mathbb{E}\, h(X) - h; \psi \rangle.$$

To conclude let us deal with the right-hand-side of (5.30). First, note that $\mathcal{A}(P_t^\nu(h)) = \mathcal{A}(P_t^\nu(\hat{h}))$. Thus,

$$\int_0^{+\infty} \langle \mathcal{A}(P_t^\nu(h)); \psi \rangle dt = \langle \int_0^{+\infty} \mathcal{A}(P_t^\nu(\hat{h})) dt; \psi \rangle$$
$$= -\langle \mathcal{A}(f_h); \psi \rangle,$$

where the interchange of integrals is justified by Fubini theorem. Then, for all $\psi \in \mathcal{S}(\mathbb{R})$,

$$\langle \mathcal{A}(f_h) + \mathbb{E}\, h(X) - h; \psi \rangle = 0.$$

Now, for $x \in \mathbb{R}$ and for $0 < \varepsilon \leq 1$, let $\psi_\varepsilon \in \mathcal{S}(\mathbb{R})$ be defined, for all $y \in \mathbb{R}$, by

$$\psi_\varepsilon(y) := \frac{1}{\sqrt{2\pi\varepsilon}} \exp\left(-\frac{(x-y)^2}{2\varepsilon^2}\right).$$

By the dominated convergence theorem and the regularity of f_h and of h, it follows that

$$\lim_{\varepsilon \to 0^+} \langle \mathcal{A}(f_h) + \mathbb{E}\, h(X) - h; \psi_\varepsilon \rangle = \mathcal{A}(f_h)(x) + \mathbb{E}\, h(X) - h(x) = 0,$$

finishing the proof of the lemma.   □

Remark 5.9 (i) *The methodology developed in this chapter can also be developed for some discrete infinitely divisible distributions such as for $X \sim \mathcal{P}(\lambda)$, a Poisson random variable with parameter $\lambda > 0$. Indeed, very classically, or from Theorem 3.1, one can infer the following Stein equation:*

$$\lambda(f(n+1) - f(n)) + (\lambda - n)f(n) = h(n) - \mathbb{E}h(X), \quad n \in \mathbb{N}, \tag{5.31}$$

with h a measurable test function such that $\mathbb{E}|h(X)| < +\infty$. Now, assume that f can be written as a discrete gradient, namely, $f(n) = g(n) - g(n - 1)$, for all $n \geq 1$. Then, (5.31) boils down to

$$\lambda(g(n + 1) - g(n) - (g(n) - g(n - 1))) + (\lambda - n)(g(n) - g(n - 1))$$
$$= h(n) - \mathbb{E}h(X), \quad n \geq 1, \quad (5.32)$$

which is the discrete analog of the continuous equation (5.14) displayed above. In particular, the left-hand side of (5.32) can be identified as the generator of the (birth–death rate process) $M/M/\infty$ queue with parameters λ and 1 (see, e.g., [8, 22, 24]). Thanks to [24, Equation 1.2], its semigroup admits the following Mehler-type representation:

$$P_t(h)(n) := \mathbb{E}h\left(\sum_{k=1}^{n} Z_k(t) + Y_t\right), \quad n \in \mathbb{N}, \, t \geq 0, \quad (5.33)$$

with h a bounded function on $\mathbb{N}$, Y_t a Poisson random variable with parameter $\lambda(1 - e^{-t})$ and $(Z_i(t))_{1 \leq i \leq n}$ iid Bernoulli random variables with parameter e^{-t} independent of Y_t, for all $t > 0$. Indeed, using once again Fourier techniques to express the generator of the $M/M/\infty$ queue as a pseudo-differential operator, it is readily seen that $(P_t)_{t>0}$ given by (5.33) is a semigroup of operators whose infinitesimal generator is given by the left-hand side of (5.32). To end this analogy, note that a notion of discrete self-decomposability has been introduced and studied (see, e.g., [87, Chap. V, Sect. 4]) to circumvent the fact that discrete random variables are not stable with respect to multiplication by a scalar. Now, a discrete random variable Y taking nonnegative values is called discrete self-decomposable, if for any $\gamma \in (0, 1)$,

$$Y =_d \gamma \circ Y + Y_\gamma, \quad (5.34)$$

where $\gamma \circ Y =_d \sum_{k=1}^{Y} Z_i(\gamma)$, with $(Z_i(\gamma))_{i \geq 1}$ a sequence of iid Bernoulli random variable with parameter γ independent of Y, and where Y_γ is a random variable independent of $\gamma \circ Y$. Discrete self-decomposable distributions are, in particular, infinitely divisible (see [87, Chap. V, Sect. 4, Theorem 4.7]) and, for instance, both the Poisson and the negative binomial distribution are discrete self-decomposable. Indeed, for any $\gamma \in (0, 1)$, the Poisson distribution satisfies the following equality in law:

$$X =_d \sum_{k=1}^{X} Z_k(\gamma) + X_\gamma, \quad (5.35)$$

with X_γ, independent of $\{X, (Z_i(\gamma))_{i \geq 1}\}$, having a Poisson distribution with parameter $\lambda(1 - \gamma)$. As previously, taking $\gamma = e^{-t}$ in (5.35) explains the Mehler representation of the semigroup of the $M/M/\infty$ queue with parameters $(\lambda, 1)$.

(ii) In a rather similar fashion, let us consider the Stein equation for the negative bino-mial distribution with parameters (r, p) as in Example 3.1 (ii). For $X \sim \mathcal{N}Bin^0(r, p)$ and for h a measurable test function such that $\mathbb{E}|h(X)| < +\infty$, then

$$(1 - p)(r + n) f(n + 1) - nf(n) = h(n) - \mathbb{E}h(X), \quad n \ge 0. \tag{5.36}$$

As before, setting $f(n) = g(n) - g(n - 1)$, (5.36) boils down to,

$$(1 - p)(r + n)(g(n + 1) - g(n)) - n(g(n) - g(n - 1)) = h(n) - \mathbb{E}h(X), \quad n \ge 1. \tag{5.37}$$

Again the left-hand side of (5.37) is the generator of an immigration–birth–death process with constant immigration rate $r(1 - p)$ and per capita birth and death rates $1 - p$ and 1 (see, e.g., [12]). Moreover, the negative binomial distribution is discrete self-decomposable since, for all $\gamma \in (0, 1)$,

$$X =_d \gamma \circ X + X_\gamma,$$

where X_γ and $\gamma \circ X$ are independent and X_γ has a probability generating function given by

$$\mathbb{E}z^{X_\gamma} = \left(\gamma + (1 - \gamma)\frac{1 - p}{1 - pz}\right)^r, \quad 0 \le z \le 1,$$

(see [87, Chap. V, Sect. 4, Example 4.6]).

(iii) To finish this chapter, let us briefly explain how self-decomposability and its discrete version, say in the Poisson case, can be combined in a single encompassing framework. Let X be a nondegenerate self-decomposable random variable with finite first moment, Lévy measure ν and characteristic function φ and let Y be a Poisson random variable with parameter $\lambda > 0$ independent of X. Then, let us define, for all $f \in \mathcal{S}(\mathbb{R})$, all $t \ge 0$, all $n \in \mathbb{N}$ and all $x \in \mathbb{R}$

$$P_t(f)(x + n) = \frac{1}{2\pi}\int_{\mathbb{R}} \mathcal{F}(f)(\xi)e^{i\xi xe^{-t}}\frac{\varphi(\xi)}{\varphi(e^{-t}\xi)}\left(e^{i\xi}e^{-t} + 1 - e^{-t}\right)^n e^{\lambda(1 - e^{-t})(e^{i\xi} - 1)}d\xi. \tag{5.38}$$

In particular, $P_t(f)(x + n)$ admits the following stochastic representation:

$$P_t(f)(x + n) = \mathbb{E}f\left(xe^{-t} + \sum_{k=1}^{n} Z_k(t) + Y_t + X_t\right), \quad x \in \mathbb{R}, n \in \mathbb{N}, t \ge 0,$$

where $\{Z_1(t), \ldots, Z_n(t), Y_t, X_t\}$ is a collection of independent random variables and where, for $1 \le i \le n$, $Z_i(t)$ is a Bernoulli random variable with parameter e^{-t}, Y_t is a Poisson random variable with parameter $\lambda(1 - e^{-t})$ and X_t has characteristic function φ_t given, for all $\xi \in \mathbb{R}$, by $\varphi_t(\xi) = \varphi(\xi)/\varphi(e^{-t}\xi)$. Moreover, from (5.38), one has, for all $n \in \mathbb{N}$ and all $x \in \mathbb{R}$,

$$\lim_{t \to +\infty} P_t(f)(x+n) = \int_{\mathbb{R}} f(x) \, (\mu_X * \mu_Y) \, (dx), \quad \lim_{t \to 0^+} P_t(f)(x+n) = f(x+n),$$

where $X \sim \mu_X$ and $Y \sim \mu_Y$. Now, using (5.38), $\lim_{t \to 0^+} (P_t(f)(x+n) - f(x+n))/t$ can be computed. Indeed, note that, for all $x \in \mathbb{R}$ and all $\xi \in \mathbb{R}$,

$$\lim_{t \to 0^+} \frac{1}{t} \left(e^{i\xi x(e^{-t}-1)} \frac{\varphi(\xi)}{\varphi(e^{-t}\xi)} - 1 \right) = -i\xi x + i\xi \mathbb{E}X + i\xi \int_{\mathbb{R}} \left(e^{iu\xi} - 1 \right) u\nu(du),$$

and, for all $n \in \mathbb{N}$ and all $\xi \in \mathbb{R}$,

$$\lim_{t \to 0^+} \frac{1}{t} \left(\left(e^{i\xi} e^{-t} + 1 - e^{-t} \right)^n e^{\lambda(1-e^{-t})(e^{i\xi}-1)} e^{-in\xi} - 1 \right) = n \left(e^{-i\xi} - 1 \right) + \lambda \left(e^{i\xi} - 1 \right).$$

Thus, for all $f \in S(\mathbb{R})$, all $n \in \mathbb{N}$ and all $x \in \mathbb{R}$,

$$\lim_{t \to 0^+} \frac{P_t(f)(x+n) - f(x+n)}{t} = (\mathbb{E}X - x)f'(x+n) + \int_{\mathbb{R}} \left(f'(x+n+u) - f'(x+n) \right) u\nu(du)$$

$$+ n(f(x+n-1) - f(x+n)) + \lambda(f(x+n+1) - f(x+n)).$$
$$(5.39)$$

From (5.39), one can infer the following Stein equation associated with the law of $X + Y$, for all $n \in \mathbb{N}$ and all $x \in \mathbb{R}$:

$$(\mathbb{E}X - x)f'_h(x+n) + \int_{\mathbb{R}} \left(f'_h(x+n+u) - f'_h(x+n) \right) u\nu(du)$$

$$+ n(f_h(x+n-1) - f_h(x+n)) + \lambda(f_h(x+n+1) - f_h(x+n)) = h(x+n) - \mathbb{E}h(X+Y),$$

where h is a suitable test function such that $\mathbb{E}\,|h(X+Y)| < +\infty$. In particular, one can retrieve the Stein's equation associated with the law of X (resp. the law of Y) by taking $\lambda = 0$ and $n = 0$ (resp. $\mathbb{E}X = 0$, $x = 0$, $\nu = 0$) and thus recovering (5.14) (resp. (5.32)).

Chapter 6
Applications to Sums of Independent Random Variables

It is well known that self-decomposable laws naturally appear as limiting laws for, rather general, sums of independent random variables (see [59, 62]). Indeed, let $(Z_k)_{k \geq 1}$ be a sequence of independent random variables, let $(b_n)_{n \geq 1}$ be a sequence of strictly positive reals, let $(c_n)_{n \geq 1}$ be a sequence of reals, and finally let

$$S_n := b_n \sum_{k=1}^{n} Z_k + c_n. \tag{6.1}$$

Recall that if $\{b_n Z_k, \ k = 1, ..., n, \ n \geq 1\}$ is a null array, i.e., if for all $\varepsilon > 0$,

$$\lim_{n \to +\infty} \max_{1 \leq k \leq n} \mathbb{P}\left(|b_n Z_k| > \varepsilon\right) = 0, \tag{6.2}$$

then whenever $(S_n)_{n \geq 1}$ converges in law, its limit has to be self-decomposable (see [84, Theorem 15.3]). (In case the limit is nondegenerate, then necessarily $b_n \to 0$ and $b_{n+1}/b_n \to 1$, as $n \to +\infty$.) Conversely, if X is self-decomposable, one can always find $(Z_k)_{k \geq 1}$, $(b_n)_{n \geq 1}$ and $(c_n)_{n \geq 1}$ as above, also satisfying (6.2) and such that $(S_n)_{n \geq 1}$ converges in law toward X. In the sequel, we quantitatively revisit this result with the help of the Stein methodology developed in the previous chapters. First, we need the following straightforward lemma.

Lemma 6.1 *Let* $(Z_k)_{k \geq 1}$, $(b_n)_{n \geq 1}$ *and* $(c_n)_{n \geq 1}$ *be as above, let* $\mathbb{E}|Z_k| < \infty$ *for all* $k \geq 1$ *and let* $f \in Lip(\mathbb{R})$. *Then, for all* $n \geq 1$,

$$\mathbb{E}S_n f'(S_n) = \left(c_n + b_n \sum_{k=1}^{n} \mathbb{E}Z_k\right) \mathbb{E}f'(S_n) + b_n \sum_{k=1}^{n} \mathbb{E}\mathbb{1}_{b_n|Z_k| \leq N} \tilde{Z}_k(f'(S_n) - f'(S_{n,k}))$$

$$+ b_n \sum_{k=1}^{n} \mathbb{E}\mathbb{1}_{b_n|Z_k| > N} \tilde{Z}_k(f'(S_n) - f'(S_{n,k})), \tag{6.3}$$

© The Author(s), under exclusive license to Springer Nature Switzerland AG 2019
B. Arras and C. Houdré, *On Stein's Method for Infinitely Divisible Laws with Finite First Moment*, SpringerBriefs in Probability and Mathematical Statistics,
https://doi.org/10.1007/978-3-030-15017-4_6

where $S_{n,k} = S_n - b_n Z_k$, $\tilde{Z}_k = Z_k - \mathbb{E}Z_k$ and $N \geq 1$. Moreover, if f is twice continuously differentiable with $\|f'\|_\infty < +\infty$ and $\|f''\|_\infty < +\infty$, then

$$\mathbb{E}S_n f'(S_n) = \left(c_n + b_n \sum_{k=1}^{n} \mathbb{E}Z_k\right) \mathbb{E}f'(S_n) + b_n \sum_{k=1}^{n} \mathbb{E}\mathbb{1}_{b_n|Z_k|>N} Z_k (f'(S_n) - f'(S_{n,k}))$$

$$+ \sum_{k=1}^{n} \int_{-\infty}^{+\infty} \mathbb{E}K_k(t,N) f''(S_{n,k}+t)dt$$

$$- b_n \sum_{k=1}^{n} \mathbb{E}Z_k \mathbb{E}(f'(S_n) - f'(S_{n,k})), \quad (6.4)$$

where

$$K_k(t,N) = \mathbb{E}b_n Z_k \mathbb{1}_{b_n|Z_k|\leq N}(\mathbb{1}_{0\leq t\leq b_n Z_k} - \mathbb{1}_{b_n Z_k\leq t\leq 0}). \quad (6.5)$$

Proof For f Lipschitz, the first part of the lemma follows from

$$\mathbb{E}\tilde{Z}_k f'(S_n) = \mathbb{E}\tilde{Z}_k(f'(S_n) - f'(S_{n,k})), \quad (6.6)$$

which is valid for all $k \geq 1$, since $\tilde{Z}_k$ and $S_{n,k}$ are independent and since $\mathbb{E}\tilde{Z}_k = 0$. For f twice continuously differentiable with $\|f'\|_\infty < +\infty$ and $\|f''\|_\infty < +\infty$, the result follows by computations similar to the proof of [94, Lemma 4.5] using also Lemma 5.5 above. □

In the next theorem, $X \sim ID(b, 0, \nu)$ is nondegenerate, self-decomposable, and such that $\mathbb{E}|X| < +\infty$. Further, $(S_n)_{n\geq 1}$ is as in (6.1) with $\mathbb{E}|Z_k| < +\infty$, for all $k \geq 1$. From the results of the previous chapter (Lemma 5.8), for any $h \in C_c^\infty(\mathbb{R})$ with $\|h\|_\infty \leq 1$, $\|h'\|_\infty \leq 1$ and $\|h''\|_\infty \leq 1$,

$$|\mathbb{E}h(S_n) - \mathbb{E}h(X)| = \left|\mathbb{E}\left((\mathbb{E}X - S_n)f_h'(S_n) + \int_{-\infty}^{+\infty}(f_h'(S_n+u) - f_h'(S_n))u\nu(du)\right)\right|. \quad (6.7)$$

The next results upperbound (6.7).

Theorem 6.2 *(i) Let $\int_{|u|\leq 1}|u|\nu(du) < +\infty$. Then, for $n, N \geq 1$,*

$$d_{W_2}(S_n, X) \leq \left|\mathbb{E}X - \left(c_n + b_n \sum_{k=1}^{n}\mathbb{E}Z_k\right)\right| + \frac{b_n}{n}\left(\sum_{k=1}^{n}\mathbb{E}|Z_k|\right)\left(\int_{-\infty}^{+\infty}|u|\nu(du)\right)$$

$$+ \frac{1}{2}b_n^2 \sum_{k=1}^{n}|\mathbb{E}Z_k|\mathbb{E}|Z_k| + 2\int_{|u|>N}|u|\nu(du) + 2b_n\sum_{k=1}^{n}\mathbb{E}|Z_k|\mathbb{1}_{|b_n Z_k|>N}$$

$$+ \frac{1}{2} \sum_{k=1}^{n} \int_{-N}^{+N} \left| \frac{K_\nu(t, N)}{n} - K_k(t, N) \right| dt. \tag{6.8}$$

(ii) Let $\int_{|u| \leq 1} |u| \nu(du) = +\infty$ and let there exist $\gamma > 0$, $\beta > 0$ and $C_1 > 0$, $C_2 > 0$ such that for $R > 0$

$$\int_{|u| > R} |u| \nu(du) \leq \frac{C_1}{R^\gamma}, \qquad \int_{|u| \leq R} |u|^2 \nu(du) \leq C_2 R^\beta. \tag{6.9}$$

Then, for $n, N \geq 1$,

$$d_{W_2}(S_n, X) \leq \left| \mathbb{E} X - \left(c_n + b_n \sum_{k=1}^{n} \mathbb{E} Z_k \right) \right| + C_{\gamma,\beta} \frac{b_n^{\frac{\beta}{\beta+\gamma}}}{n} \left(\sum_{k=1}^{n} \mathbb{E} |Z_k|^{\frac{\beta}{\beta+\gamma}} \right)$$

$$+ \frac{1}{2} b_n^2 \sum_{k=1}^{n} |\mathbb{E} Z_k| \mathbb{E} |Z_k| + 2 \int_{|u| > N} |u| \nu(du) + 2 b_n \sum_{k=1}^{n} \mathbb{E} |Z_k| \mathbb{1}_{|b_n Z_k| > N}$$

$$+ \frac{1}{2} \sum_{k=1}^{n} \int_{-N}^{+N} \left| \frac{K_\nu(t, N)}{n} - K_k(t, N) \right| dt,$$

for some $C_{\gamma,\beta} > 0$ only depending on γ and β.

Proof Let us start with the proof of (i). Assume that $\int_{|u| \leq 1} |u| \nu(du) < +\infty$. Let $h \in C_c^\infty(\mathbb{R})$ be such that $\|h\|_\infty \leq 1$, $\|h'\|_\infty \leq 1$ and $\|h''\|_\infty \leq 1$. Then,

$$|\mathbb{E} h(S_n) - \mathbb{E} h(X)| = \left| \mathbb{E} \left((\mathbb{E} X - S_n) f_h'(S_n) + \int_{-\infty}^{+\infty} \left(f_h'(S_n + u) - f_h'(S_n) \right) u \nu(du) \right) \right|$$

$$\leq \left| \mathbb{E} \left(\mathbb{E} X - \left(c_n + b_n \sum_{k=1}^{n} \mathbb{E} Z_k \right) \right) f_h'(S_n) \right|$$

$$+ \left| \mathbb{E} b_n \sum_{k=1}^{n} \tilde{Z}_k (f_h'(S_{n,k} + Z_k b_n) - f_h'(S_{n,k})) - \mathcal{T}(f_h)(S_n) \right|$$

$$\leq \left| \mathbb{E} X - \left(c_n + b_n \sum_{k=1}^{n} \mathbb{E} Z_k \right) \right| + \left| \mathbb{E} \frac{1}{n} \sum_{k=1}^{n} \mathcal{T}(f_h)(S_{n,k}) - \mathcal{T}(f_h)(S_n) \right|$$

$$+ \left| \mathbb{E} b_n \sum_{k=1}^{n} \tilde{Z}_k (f_h'(S_{n,k} + Z_k b_n) - f_h'(S_{n,k})) - \frac{1}{n} \sum_{k=1}^{n} \mathcal{T}(f_h)(S_{n,k}) \right|$$

$$\leq \left| \mathbb{E} X - \left(c_n + b_n \sum_{k=1}^{n} \mathbb{E} Z_k \right) \right| + \frac{b_n}{n} \left(\int_{-\infty}^{+\infty} |u| \nu(du) \right) \sum_{k=1}^{n} \mathbb{E} |Z_k|$$

$$+ \left| \mathbb{E} b_n \sum_{k=1}^{n} \tilde{Z}_k (f_h'(S_{n,k} + Z_k b_n) - f_h'(S_{n,k})) - \frac{1}{n} \sum_{k=1}^{n} \mathcal{T}(f_h)(S_{n,k}) \right|, \tag{6.10}$$

where we have successively used Lemma 5.4 and Proposition 5.6 (i). Next, we need to bound

$$I := \left| \mathbb{E}b_n \sum_{k=1}^{n} \tilde{Z}_k(f_h'(S_{n,k} + Z_k b_n) - f_h'(S_{n,k})) - \frac{1}{n} \sum_{k=1}^{n} \mathcal{T}(f_h)(S_{n,k}) \right|.$$

At first,

$$I \le \left| \mathbb{E} b_n \sum_{k=1}^{n} \mathbb{E}Z_k(f_h'(S_{n,k} + Z_k b_n) - f_h'(S_{n,k})) \right|$$

$$+ \left| \mathbb{E} b_n \sum_{k=1}^{n} Z_k(f_h'(S_{n,k} + Z_k b_n) - f_h'(S_{n,k})) - \frac{1}{n} \sum_{k=1}^{n} \mathcal{T}(f_h)(S_{n,k}) \right|$$

$$\le \frac{1}{2} b_n^2 \sum_{k=1}^{n} |\mathbb{E}Z_k| \mathbb{E}|Z_k|$$

$$+ \left| \mathbb{E}b_n \sum_{k=1}^{n} Z_k(f_h'(S_{n,k} + Z_k b_n) - f_h'(S_{n,k})) - \frac{1}{n} \sum_{k=1}^{n} \mathcal{T}(f_h)(S_{n,k}) \right|.$$

Let $N \ge 1$. Now, from Lemma 5.5,

$$\frac{1}{n} \sum_{k=1}^{n} \mathcal{T}(f_h)(S_{n,k}) = \frac{1}{n} \sum_{k=1}^{n} \left(\int_{-N}^{+N} K_\nu(t, N) f_h''(S_{n,k} + t) dt + R_N(S_{n,k}) \right),$$

and moreover,

$$\frac{1}{n} \sum_{k=1}^{n} \mathbb{E}|R_N(S_{n,k})| \le 2 \int_{|u|>N} |u| \nu(du).$$

Therefore,

$$I \le \frac{1}{2} b_n^2 \sum_{k=1}^{n} |\mathbb{E}Z_k| \mathbb{E}|Z_k| + 2 \int_{|u|>N} |u| \nu(du) + \left| \mathbb{E}b_n \sum_{k=1}^{n} Z_k(f_h'(S_{n,k} + Z_k b_n) - f_h'(S_{n,k})) \right.$$

$$\left. - \frac{1}{n} \sum_{k=1}^{n} \int_{-N}^{+N} K_\nu(t, N) f_h''(S_{n,k} + t) dt \right|,$$

and from Lemma 6.1,

$$\mathbb{E}b_n \sum_{k=1}^{n} Z_k(f_h'(S_{n,k} + Z_k b_n) - f_h'(S_{n,k})) = b_n \sum_{k=1}^{n} \mathbb{E}\mathbb{1}_{b_n |Z_k| > N} Z_k(f_h'(S_{n,k} + Z_k b_n) - f_h'(S_{n,k}))$$

$$+ \sum_{k=1}^{n} \int_{-\infty}^{+\infty} \mathbb{E}K_k(t, N) f_h''(S_{n,k} + t) dt.$$

Then,

$$
I \le \frac{1}{2} b_n^2 \sum_{k=1}^{n} |\mathbb{E} Z_k| \mathbb{E} |Z_k| + 2 \int_{|u|>N} |u| \nu(du) + 2 b_n \sum_{k=1}^{n} \mathbb{E} |Z_k| \mathbb{1}_{|b_n Z_k| > N}
$$
$$
+ \frac{1}{2} \sum_{k=1}^{n} \int_{-N}^{+N} \left| \frac{K_\nu(t, N)}{n} - K_k(t, N) \right| dt. \tag{6.11}
$$

Combining (6.10) and (6.11) proves part (i). For the proof of (ii), proceed in a similar way. □

Remark 6.3 *(i) Using Lemma A.4, it is possible to transfer the bounds on $d_{W_2}(S_n, X)$ obtained above to bounds on $d_{W_1}(S_n, X)$.*
(ii) The approach just presented generalizes the methodology developed for the symmetric α-stable distribution in [94]. Note that in this case, due to the regularizing properties of the probability transition density function of the one-dimensional α-stable Lévy process, it is possible to obtain quantitative upper bounds in Wasserstein-1 distance. At the level of the semigroup solution to the Stein equation, this can be viewed by a gain of one order of differentiability with a test function h only Lipschitz. In this regard, Theorem 6.2 can be compared with Theorem 2.1 of [94] (from which several quantitative convergence results follow).
(iii) As indicated in [94], the quantity

$$
\frac{1}{2} \sum_{k=1}^{n} \int_{-N}^{+N} \left| \frac{K_\nu(t, N)}{n} - K_k(t, N) \right| dt
$$

might be seen as the L^1-analog of the L^2-Stein discrepancy considered, e.g., in [25, 26, 61, 71]. However, the technique developed in the current chapter is much more reminiscent of the K-function approach exposed, e.g., in [31, Sect. 2.3.1] when dealing with sums of independent summands in a Gaussian setting.
(iv) To illustrate the abstract bounds obtained in the previous theorem, let us consider a canonical example associated with nondegenerate self-decomposable laws (see [84, proof of Theorem 15.3, (ii)]). Let $X \sim ID(b, 0, \nu)$ be nondegenerate self-decomposable with $\mathbb{E}|X| < \infty$, $X \ge 0$ a.s. and let φ be its characteristic function. For all $n \ge 1$, set $c_n = 0$ and $b_n = 1/n$. Further, let $(Z_m)_{m \ge 1}$ be independent random variables defined via their characteristic functions, given for all $m \ge 1$ and $\xi \in \mathbb{R}$ by

$$
\varphi_m(\xi) := \frac{\varphi((m+1)\xi)}{\varphi(m\xi)} = \exp\left(i\xi \mathbb{E}X + \int_0^{+\infty} (e^{iu\xi} - 1 - iu\xi) \frac{k_m(u)}{u} du \right), \tag{6.12}
$$

with $k_m(u) = k(u/(m+1)) - k(u/m)$ and k given by $\nu(du)/du = k(u)/|u|$, thanks to [84, Corollary 15.11]. Then, the sequence $(S_n)_{n \ge 1}$ defined by (6.1)

converges in distribution to X. Moreover, it is possible to extract, for this exam-
ple, explicit rates of convergence for some of the terms present on the right-hand
side of (6.8). First, note that $\mathbb{E}X - (c_n + b_n \sum_{m=1}^{n} \mathbb{E}Z_m) = 0$, for all $n \geq 1$. Next,
consider the terms defined, for all $n \geq 1$, by

$$(II) := \frac{b_n}{n} \left(\sum_{m=1}^{n} \mathbb{E}|Z_m| \right) \left(\int_{-\infty}^{+\infty} |u|\nu(du) \right), \quad (III) := \frac{1}{2}b_n^2 \sum_{m=1}^{n} |\mathbb{E}Z_m|\mathbb{E}|Z_m|.$$

Thanks to [84, Theorem 24.11], $Z_m \geq 0$ a.s. for all $m \geq 1$, thus for all $n \geq 1$

$$(II) \leq \frac{\mathbb{E}X}{n} \left(\int_{0}^{+\infty} u\nu(du) \right), \quad (III) \leq \frac{(\mathbb{E}X)^2}{2n}.$$

The last three terms on the right-hand side of (6.8) depend, respectively, on the tail
properties of the Lévy measure ν of X, on $\mathbb{P}(Z_m \geq x)$, as $x \to +\infty$, and on a refined
analysis of the L^1-Stein discrepancy combined with a good choice of the truncation
parameter N.

Next, we consider the important case where $(Z_k)_{k\geq 1}$ is a sequence of independent
and identically distributed random variables such that $\mathbb{E}|Z_k| < +\infty$, $k \geq 1$. In this
situation, from [84, Theorem 15.7], the limiting self-decomposable law is actually
stable. Recall that an infinitely divisible probability measure, μ, is stable if, for any
$a > 0$, there exist $b > 0$ and $c \in \mathbb{R}$ such that, for all $t \in \mathbb{R}$,

$$\varphi(t)^a = \varphi(bt)e^{ict},$$

where φ is the characteristic function of μ. Also, by [84, Theorem 14.3(ii)], any
nondegenerate stable distribution with index $\alpha \in (0, 2)$ has a Lévy measure given by
(2.10). Note finally that, when ν is given by (2.10) with $\alpha \in (1, 2)$, $\int_{|u|\leq 1} |u|\nu(du) = +\infty$ and that, for any $R > 0$,

$$\int_{|u|>R} |u|\nu(du) = \frac{c_1 + c_2}{\alpha - 1} \frac{1}{R^{\alpha-1}}, \quad \int_{|u|\leq R} |u|^2\nu(du) = \frac{c_1 + c_2}{2 - \alpha} R^{2-\alpha}.$$

Corollary 6.4 *Let X be a nondegenerate stable random variable with index*
$\alpha \in (1, 2)$ *and Lévy measure given by (2.10). Let $(S_n)_{n\geq 1}$ be as in (6.1) with $(Z_k)_{k\geq 1}$*
independent and identically distributed and with $\mathbb{E}|Z_1| < +\infty$. Then, for $n, N \geq 1$,

$$d_{W_2}(S_n, X) \leq |\mathbb{E}X - (c_n + nb_n\mathbb{E}Z_1)| + C_\alpha(b_n)^{2-\alpha}\mathbb{E}|Z_1|^{2-\alpha} + \frac{n}{2}b_n^2|\mathbb{E}Z_1|\mathbb{E}|Z_1|$$

$$+ 2\frac{c_1 + c_2}{\alpha - 1} \frac{1}{N^{\alpha-1}} + 2nb_n\mathbb{E}|Z_1|\mathbb{1}_{|b_n Z_1|>N} + \frac{1}{2}\int_{-N}^{+N} |K_\nu(t, N) - nK_1(t, N)|\, dt,$$

$$(6.13)$$

where $C_\alpha > 0$ only depends on α, c_1 and c_2, and where, for all $t \in \mathbb{R}$, $t \neq 0$ and
$N \geq 1, n \geq 1$,

$$K_\nu(t, N) = c_1 \mathbb{1}_{[0,N]}(t)\frac{t^{1-\alpha} - N^{1-\alpha}}{\alpha - 1} + c_2 \mathbb{1}_{[-N,0]}(t)\frac{N^{1-\alpha} - (-t)^{1-\alpha}}{1 - \alpha},$$

$$K_1(t, N) = \mathbb{E}b_n Z_1 \mathbb{1}_{b_n|Z_1|\leq N}(\mathbb{1}_{0\leq t\leq b_n Z_1} - \mathbb{1}_{b_n Z_1 \leq t \leq 0}).$$

Proof This is a direct application of Theorem 6.2 (ii) ($\beta = 2 - \alpha$ and $\gamma = \alpha - 1$) together with the fact that the random variables Z_k are identically distributed. Moreover, thanks to (2.10) with $\alpha \in (1, 2)$ and to Lemma 5.5, for all $t \in \mathbb{R}$, $t \neq 0$, and for all $N \geq 1$

$$K_\nu(t, N) = c_1 \mathbb{1}_{[0,N]}(t)\frac{t^{1-\alpha} - N^{1-\alpha}}{\alpha - 1} + c_2 \mathbb{1}_{[-N,0]}(t)\frac{N^{1-\alpha} - (-t)^{1-\alpha}}{1 - \alpha}.$$

This concludes the proof of the corollary.   □

Remark 6.5 *Assuming further properties of the tails behavior of the law of Z_1 as done, for example, in [94, Theorem 2.6 and Corollary 2.7], it is possible to extract explicit rates of convergence from the right-hand side of (6.13). For more details, the reader is referred to the proofs of [94, Theorem 2.6 and Corollary 2.7].*

Finally, as ultimate example, let now $(Z_{n,k})_{1\leq k\leq n, n\geq 1}$ be a null array (see, e.g., [84, Definition 9.2]), namely, for each fixed $n \geq 1$, $Z_{n,1}, \ldots, Z_{n,n}$ are independent random variables and, for all $\varepsilon > 0$,

$$\lim_{n\to+\infty} \max_{1\leq k\leq n} \mathbb{P}\left(|Z_{n,k}| > \varepsilon\right) = 0. \tag{6.14}$$

Let $(S_n)_{n\geq 1}$ be the sequence of row sums associated with this triangular array, namely, for all $n \geq 1$ let

$$S_n = \sum_{k=1}^n Z_{n,k}. \tag{6.15}$$

A classical result of Khintchine (see, e.g., [84, Theorem 9.3]) asserts that if for some $c_n \in \mathbb{R}$, $n \geq 1$, $S_n + c_n$ converges in distribution to X, then X is infinitely divisible. Thus, by a straightforward adaptation of the proofs of Lemma 6.1 and of Theorem 6.2, the following result for sums from null arrays holds true.

Theorem 6.6 *Let $X \sim ID(b, 0, \nu)$ be nondegenerate, self-decomposable and such that $\mathbb{E}|X| < \infty$. Let $(S_n)_{n\geq 1}$ be given by (6.15) with $\mathbb{E}|Z_{n,k}| < \infty$, for all $n \geq 1$, for all $k = 1, \ldots, n$ and let $c_n \in \mathbb{R}$, for all $n \geq 1$.*
(i) Let $\int_{|u|\leq 1} |u|\nu(du) < +\infty$. Then, for $n, N \geq 1$,

$$d_{W_2}(S_n + c_n, X) \le \left| \mathbb{E}X - \left(c_n + \sum_{k=1}^{n} \mathbb{E}Z_{n,k}\right) \right| + \frac{1}{n}\left(\sum_{k=1}^{n} \mathbb{E}|Z_{n,k}|\right)\left(\int_{-\infty}^{+\infty} |u|\nu(du)\right)$$

$$+ \frac{1}{2}\sum_{k=1}^{n} |\mathbb{E}Z_{n,k}|\mathbb{E}|Z_{n,k}| + 2\int_{|u|>N} |u|\nu(du) + 2\sum_{k=1}^{n} \mathbb{E}|Z_{n,k}|\mathbb{1}_{|Z_{n,k}|>N}$$

$$+ \frac{1}{2}\sum_{k=1}^{n} \int_{-N}^{+N} \left| \frac{K_\nu(t,N)}{n} - K_k(t,N) \right| dt,$$

where, for all $t \in \mathbb{R}$, for all $N \ge 1$ and for all $1 \le k \le n$

$$K_k(t,N) = \mathbb{E}Z_{n,k}\mathbb{1}_{|Z_{n,k}|\le N}(\mathbb{1}_{0\le t\le Z_{n,k}} - \mathbb{1}_{Z_{n,k}\le t\le 0}).$$

(ii) Let $\int_{|u|\le 1} |u|\nu(du) = +\infty$ and let there exist $\gamma > 0$, $\beta > 0$ and $C_1 > 0, C_2 > 0$ such that for any $R > 0$

$$\int_{|u|>R} |u|\nu(du) \le \frac{C_1}{R^\gamma}, \qquad \int_{|u|\le R} |u|^2\nu(du) \le C_2 R^\beta.$$

Then, for $n, N \ge 1$,

$$d_{W_2}(S_n + c_n, X) \le \left| \mathbb{E}X - \left(c_n + \sum_{k=1}^{n} \mathbb{E}Z_{n,k}\right) \right| + \frac{C_{\gamma,\beta}}{n}\left(\sum_{k=1}^{n} \mathbb{E}|Z_{n,k}|^{\frac{\beta}{\beta+\gamma}}\right)$$

$$+ \frac{1}{2}\sum_{k=1}^{n} |\mathbb{E}Z_{n,k}|\mathbb{E}|Z_{n,k}| + 2\int_{|u|>N} |u|\nu(du) + 2\sum_{k=1}^{n} \mathbb{E}|Z_{n,k}|\mathbb{1}_{|Z_{n,k}|>N}$$

$$+ \frac{1}{2}\sum_{k=1}^{n} \int_{-N}^{+N} \left| \frac{K_\nu(t,N)}{n} - K_k(t,N) \right| dt,$$

for some $C_{\gamma,\beta} > 0$ only depending on γ and β.

To conclude this chapter, let us present some concrete examples from extreme value theory and number theory for which explicit rates of convergence can be found. To start with extreme value theory, let $(Y_k)_{k\ge 1}$ be a sequence of i.i.d. exponential random variables with parameter 1. Then, for any $n \ge 1$, set,

$$S_n = \sum_{k=1}^{n} \frac{Y_k}{k} - \ln n. \tag{6.16}$$

As it is well known, $\max_{1\le k\le n} Y_k =_d \sum_{k=1}^{n} k^{-1}Y_k$ and S_n converges in distribution to a Gumbel random variable X, whose distribution function is given by $F(x) := \exp(-\exp(-x))$, for all $x \in \mathbb{R}$ (see [87, Chap. IV, Example 11.1]). Moreover, the Lévy–Khintchine representation of the characteristic function of X is given by

$$\varphi(t) = \exp\left(it\gamma + \int_0^{+\infty} \left(e^{itu} - 1 - itu \right) \frac{e^{-u}}{u(1 - e^{-u})} du \right), \quad t \in \mathbb{R},$$

where γ is the Euler constant (see [87, Chap. IV, Example 11.10]), and so X is a nondegenerate self-decomposable random variable with finite first moment. Note also that

$$\int_0^1 u\nu(du) = +\infty \quad \text{and} \quad \int_0^{+\infty} u^2\nu(du) = \frac{\pi^2}{6},$$

where ν is the Lévy measure of X. The following result provides an explicit rate of convergence which is reminiscent of the one contained in [51].

Theorem 6.7 *Let $(S_n)_{n\geq 1}$ be defined by (6.16). Let X be a Gumbel random variable with distribution function $F(x) = \exp\left(-\exp(-x) \right)$, for $x \in \mathbb{R}$. Then, for all $n \geq 1$*

$$d_{W_2}(S_n, X) \leq \frac{C}{n}$$

for some $C > 0$ independent of n.

Proof Let $h \in C_c^\infty(\mathbb{R})$ be such that $\|h\|_\infty \leq 1$, $\|h'\|_\infty \leq 1$ and $\|h''\|_\infty \leq 1$ and let f_h be given by Lemma 5.8. Then, for all $n \geq 1$

$$|\mathbb{E}h(S_n) - \mathbb{E}h(X)| = \left| \mathbb{E}(\gamma - S_n)f_h'(S_n) + \int_0^{+\infty} \left(f_h'(S_n + u) - f_h'(S_n) \right) \frac{e^{-u}}{(1 - e^{-u})} du \right|$$

$$\leq |\gamma - \mathbb{E}S_n| + \left| \mathbb{E}(\mathbb{E}S_n - S_n)f_h'(S_n) \right.$$

$$\left. + \int_0^{+\infty} \left(f_h'(S_n + u) - f_h'(S_n) \right) \frac{e^{-u}}{(1 - e^{-u})} du \right|.$$

First, note that, for all $n \geq 1$

$$|\gamma - \mathbb{E}S_n| = \left| \gamma + \ln n - \sum_{k=1}^n \frac{1}{k} \right| \leq \frac{C_1}{n}$$

for some $C_1 > 0$ not depending on n. Set $S_{n,k} = S_n - k^{-1}Y_k$, for $n \geq 1$ and $1 \leq k \leq n$. Now, for all $n \geq 1$,

$$\mathbb{E}S_n f_h'(S_n) = \sum_{k=1}^n \frac{1}{k} \mathbb{E}Y_k f_h'(S_{n,k} + k^{-1}Y_k) - \ln n \, \mathbb{E}f_h'(S_n),$$

$$= \sum_{k=1}^n \frac{1}{k} \int_0^{+\infty} e^{-u} \mathbb{E}f_h'(S_n + k^{-1}u) du - \ln n \, \mathbb{E}f_h'(S_n),$$

using that Y_k is independent of $S_{n,k}$ and Theorem 3.1 applied to Y_k. Then, for all $n \geq 1$

$$\left| \mathbb{E}(\mathbb{E}S_n - S_n) f_h'(S_n) + \int_0^{+\infty} \left(f_h'(S_n + u) - f_h'(S_n) \right) \frac{e^{-u}}{(1 - e^{-u})} du \right| = \left| \mathbb{E} \sum_{k=1}^n \frac{1}{k} \int_0^{+\infty} \left(f_h'(S_n) \right. \right.$$

$$\left. - f_h'(S_n + k^{-1}u) \right) e^{-u} du + \int_0^{+\infty} \left(f_h'(S_n + u) - f_h'(S_n) \right) \frac{e^{-u}}{(1 - e^{-u})} du \right|$$

$$= \left| \mathbb{E} \int_0^{+\infty} \left(f_h'(S_n) - f_h'(S_n + u) \right) \frac{1 - e^{-nu}}{e^u - 1} du + \int_0^{+\infty} \left(f_h'(S_n + u) \right. \right.$$

$$\left. - f_h'(S_n) \right) \frac{1}{(e^u - 1)} du \right|$$

$$= \left| \mathbb{E} \int_0^{+\infty} \left(f_h'(S_n + u) - f_h'(S_n) \right) \frac{e^{-nu}}{e^u - 1} du \right| \leq \frac{1}{2} \int_0^{+\infty} \frac{e^{-nu}}{e^u - 1} u \, du.$$

Next, [74, Formula 25.11.25], for all $n \geq 1$

$$\int_0^{+\infty} \frac{e^{-nu}}{e^u - 1} u \, du = \zeta(2, n+1),$$

where ζ is the Hurwitz zeta function defined by $\zeta(2, n+1) = \sum_{k=0}^{+\infty} (k + n + 1)^{-2}$, for all $n \geq 1$. The asymptotic expansion [74, 25.11.43] concludes the proof. $\qquad \square$

The next example deals with the fluctuations of a probabilistic model on certain type of integers which was first introduced in [23] and further studied in [4, 17]. Let $(p_j)_{j \geq 1}$ be an enumeration in increasing order of the prime numbers with $p_1 = 2$. Let $(X_k)_{k \geq 1}$ be a sequence of independent Bernoulli random variables such that, for each $k \geq 1$, $\mathbb{P}(X_k = 1) = 1/(1 + p_k)$ and $\mathbb{P}(X_k = 0) = p_k/(1 + p_k)$. For each $n \geq 1$, let now S_n be given by

$$S_n = \frac{1}{\ln p_n} \sum_{k=1}^n (\ln p_k) X_k. \tag{6.17}$$

By [23, Theorem 1], S_n converges in law, as n tends to infinity, toward a Dickman distributed random variable with parameter $\theta = 1$. The next theorem refines this result. Its proof rests in part upon a technical lemma which is contained in [4] but, for the sake of completeness, a proof is also provided in the Appendix.

Theorem 6.8 *Let $(S_n)_{n \geq 1}$ be defined by (6.17). Let X be a Dickman distributed random variable with parameter $\theta = 1$. Then, for all $n \geq 2$*

$$d_{W_2}(S_n, X) \leq \frac{C}{\ln n},$$

for some $C > 0$ independent of n.

Proof The beginning of the proof is similar to the one of Theorem 6.2 (i), noting that the Lévy measure of the Dickman distribution is given by $\nu(du) = \mathbb{1}_{(0,1)}(u) du/u$.

Let $h \in C_c^\infty(\mathbb{R})$ be such that $\|h\|_\infty \le 1$, $\|h'\|_\infty \le 1$ and $\|h''\|_\infty \le 1$ and let f_h be given by Lemma 5.8. Then, setting $b_n = (\ln p_n)^{-1}$, $Z_k = (\ln p_k)X_k$, $\tilde{Z}_k = Z_k - \mathbb{E}Z_k$ and $c_n = 0$,

$$
|\mathbb{E}h(S_n) - \mathbb{E}h(X)| \le \left| \mathbb{E}\,X - \left(c_n + b_n \sum_{k=1}^n \mathbb{E}Z_k \right) \right| + \frac{b_n}{n} \left(\int_{-\infty}^{+\infty} |u|\nu(du) \right) \sum_{k=1}^n \mathbb{E}|Z_k|
$$

$$
+ \left| \mathbb{E}\,b_n \sum_{k=1}^n \tilde{Z}_k (f_h'(S_{n,k} + Z_k b_n) - f_h'(S_{n,k})) - \frac{1}{n} \sum_{k=1}^n \mathcal{T}(f_h)(S_{n,k}) \right|,
$$

$$
\le \left| \mathbb{E}\,X - \left(c_n + b_n \sum_{k=1}^n \mathbb{E}Z_k \right) \right| + \frac{b_n}{n} \left(\int_{-\infty}^{+\infty} |u|\nu(du) \right) \sum_{k=1}^n \mathbb{E}|Z_k| + \frac{1}{2}b_n^2 \sum_{k=1}^n |\mathbb{E}Z_k|\mathbb{E}|Z_k|
$$

$$
+ \left| \mathbb{E}b_n \sum_{k=1}^n Z_k (f_h'(S_{n,k} + Z_k b_n) - f_h'(S_{n,k})) - \frac{1}{n} \sum_{k=1}^n \mathcal{T}(f_h)(S_{n,k}) \right|. \tag{6.18}
$$

Let us begin by upper bounding the first three terms on the right-hand side of (6.18). For all $n \ge 2$, using Lemma A.7,

$$
\left| \mathbb{E}\,X - \left(c_n + b_n \sum_{k=1}^n \mathbb{E}Z_k \right) \right| \le \left| 1 - \frac{1}{\ln p_n} \sum_{k=1}^n \frac{\ln p_k}{1 + p_k} \right|,
$$
$$
\le \frac{C_1}{\ln n}
$$

for some $C_1 > 0$ independent of n. Moreover, for all $n \ge 2$

$$
\frac{b_n}{n} \left(\int_{-\infty}^{+\infty} |u|\nu(du) \right) \sum_{k=1}^n \mathbb{E}|Z_k| \le \frac{1}{n \ln p_n} \sum_{k=1}^n \frac{\ln p_k}{1 + p_k},
$$
$$
\le \frac{C_3}{n},
$$

for some $C_3 > 0$ independent of n. Similarly, for all $n \ge 2$

$$
\frac{1}{2}b_n^2 \sum_{k=1}^n |\mathbb{E}Z_k|\mathbb{E}|Z_k| = \frac{1}{2(\ln p_n)^2} \sum_{k=1}^n \left(\frac{\ln p_k}{(1 + p_k)} \right)^2 = \mathcal{O}\left(\frac{1}{(\ln n)^2} \right).
$$

To conclude, let us deal with $\left| \mathbb{E}b_n \sum_{k=1}^n Z_k(f_h'(S_{n,k} + Z_k b_n) - f_h'(S_{n,k})) - \sum_{k=1}^n \mathcal{T}(f_h)(S_{n,k})/n \right|$, for all $n \ge 2$. First, note that, for all $n \ge 1$

$$
\mathbb{E}\frac{1}{n} \sum_{k=1}^n \mathcal{T}(f_h)(S_{n,k}) = \mathbb{E}\frac{1}{n} \sum_{k=1}^n \int_0^1 (f_h'(S_{n,k} + u) - f_h'(S_{n,k}))\,du = \mathbb{E}\left(f_h'(S_{n,J} + U) - f_h'(S_{n,J}) \right),
$$

where U and J are independent random variables, independent of $(X_k)_{k\ge 1}$, and uniformly distributed on $[0, 1]$ and on $\{1, \dots, n\}$, respectively. Moreover, by the very definition of I in Lemma A.7, one has

$$\mathbb{E}b_n \sum_{k=1}^{n} Z_k(f_h'(S_{n,k} + Z_k b_n) - f_h'(S_{n,k})) = \mathbb{E}S_n \mathbb{E}\left(f_h'\left(S_{n,I} + \frac{\ln p_I}{\ln p_n} \right) - f_h'(S_{n,I}) \right).$$

Then, for all $n \geq 1$

$$\left| \mathbb{E}b_n \sum_{k=1}^{n} Z_k(f_h'(S_{n,k} + Z_k b_n) - f_h'(S_{n,k})) - \frac{1}{n} \sum_{k=1}^{n} \mathcal{T}(f_h)(S_{n,k}) \right|$$

$$\leq \left| \mathbb{E}S_n \mathbb{E}\left(f_h'\left(S_{n,I} + \frac{\ln p_I}{\ln p_n} \right) - f_h'(S_{n,I}) \right) - \mathbb{E}\left(f_h'(S_{n,J} + U) - f_h'(S_{n,J}) \right) \right|,$$

$$\leq 2 \left| \mathbb{E}S_n - 1 \right| + \left| \mathbb{E}f_h'(S_{n,I}) - f_h'(S_{n,J}) \right| + \left| \mathbb{E}f_h'\left(S_{n,I} + \frac{\ln p_I}{\ln p_n} \right) - f_h'(S_{n,J} + U) \right|.$$

Let us deal with these three terms separately. First, as previously, for all $n \geq 2$,

$$|\mathbb{E}S_n - 1| \leq \frac{C_1}{\ln n}.$$

Now, for all $n \geq 2$,

$$|\mathbb{E}f_h'(S_{n,I}) - f_h'(S_{n,J})| \leq \frac{1}{2}\mathbb{E}\left| \frac{X_I \ln p_I}{\ln p_n} - \frac{X_J \ln p_J}{\ln p_n} \right| \leq \frac{1}{2}\mathbb{E}\left(\frac{X_I \ln p_I}{\ln p_n} \right) + \frac{1}{2}\mathbb{E}\left(\frac{X_J \ln p_J}{\ln p_n} \right).$$

But, by the respective definitions of I and J,

$$\mathbb{E}\left(\frac{X_I \ln p_I}{\ln p_n} \right) = \frac{1}{\mathbb{E}S_n (\ln p_n)^2} \sum_{k=1}^{n} \frac{(\ln p_k)^2}{(1 + p_k)^2} = \mathcal{O}\left(\frac{1}{(\ln n)^2} \right),$$

$$\mathbb{E}\left(\frac{X_J \ln p_J}{\ln p_n} \right) = \frac{1}{n \ln p_n} \sum_{k=1}^{n} \frac{\ln p_k}{1 + p_k} = \mathcal{O}\left(\frac{1}{n} \right).$$

Finally, for all $n \geq 1$,

$$\left| \mathbb{E}f_h'\left(S_{n,I} + \frac{\ln p_I}{\ln p_n} \right) - f_h'(S_{n,J} + U) \right| \leq \frac{1}{2}\mathbb{E}\left| U - \frac{\ln p_I}{\ln p_n} \right| + \frac{1}{2}\mathbb{E}\left(\frac{X_I \ln p_I}{\ln p_n} \right) + \frac{1}{2}\mathbb{E}\left(\frac{X_J \ln p_J}{\ln p_n} \right).$$

Using, once more, Lemma A.7 completes the proof of the upper bound.   □

We finish our manuscript by briefly addressing, among many others, three possible extensions and generalizations of our current work which will be presented elsewhere. A first possible direction of future research is to solve the Stein equation associated with general infinitely divisible distributions with finite first moment (not only the self-decomposable target laws). A second possible direction of research to which our methods are amenable is the study of extensions to multivariate (and even infinite-dimensional) settings of the results presented here. (In this vein, see [3].) A third direction would be to attempt at removing the finite first moment assumption which is present throughout our hypotheses.

Appendix

This chapter is devoted to the proof of seven technical results used in the previous chapters.

Lemma A.1 *Let $X \sim ID(b, 0, \nu)$ be self-decomposable, with characteristic function φ, and such that $\mathbb{E}|X| < \infty$. Let $x, \xi \in \mathbb{R}$. Then,*

$$\lim_{t \to 0^+} \frac{1}{t} \left(e^{i\xi x(e^{-t}-1)} \frac{\varphi(\xi)}{\varphi(e^{-t}\xi)} - 1 \right) = \left(-x + \mathbb{E}X + \int_{-\infty}^{+\infty} \left(e^{iu\xi} - 1 \right) u\nu(du) \right) (i\xi).$$

Proof Let $x, \xi \in \mathbb{R}$. First, it is clear that

$$\lim_{t \to 0^+} \frac{1}{t} (e^{i\xi x(e^{-t}-1)} - 1) = -i\xi x. \tag{A.1}$$

Next, since X is infinitely divisible with finite first moment, for all $t \geq 0$,

$$\frac{\varphi(\xi)}{\varphi(e^{-t}\xi)} = e^{i\xi \mathbb{E}X(1-e^{-t})} e^{\int_{-\infty}^{+\infty} \left(e^{iu\xi} - e^{iu\xi e^{-t}} - iu\xi(1-e^{-t}) \right) \nu(du)}.$$

Now, from (A.1),

$$\lim_{t \to 0^+} \frac{1}{t} (e^{i\xi \mathbb{E}X(1-e^{-t})} - 1) = i\xi \mathbb{E}X,$$

and moreover,

$$\lim_{t \to 0^+} e^{\int_{-\infty}^{+\infty} \left(e^{iu\xi} - e^{iu\xi e^{-t}} - iu\xi(1-e^{-t}) \right) \nu(du)} = 1.$$

So, to finish the proof, one needs to show that

$$\lim_{t\to 0^+}\frac{1}{t}\left(e^{\int_{-\infty}^{+\infty}\left(e^{iu\xi}-e^{iu\xi e^{-t}}-iu\xi(1-e^{-t})\right)\nu(du)}-1\right)=\int_{-\infty}^{+\infty}\left(e^{iu\xi}-1\right)u\nu(du)(i\xi).$$

For this purpose, let us show that

$$\lim_{t\to 0^+}\frac{1}{t}\int_{-\infty}^{+\infty}\left(e^{iu\xi}-e^{iu\xi e^{-t}}-iu\xi(1-e^{-t})\right)\nu(du)=\int_{-\infty}^{+\infty}\left(e^{iu\xi}-1\right)u\nu(du)(i\xi).$$
(A.2)

For the real part of (A.2), one wants to prove

$$\lim_{t\to 0^+}\frac{1}{t}\int_{-\infty}^{+\infty}\left(\cos(u\xi)-\cos(u\xi e^{-t})\right)\nu(du)=-\int_{-\infty}^{+\infty}\sin(u\xi)u\nu(du)\xi. \quad \text{(A.3)}$$

But, for all $u \in \mathbb{R}$,

$$\lim_{t\to 0^+}\frac{1}{t}\left(\cos(u\xi)-\cos(u\xi e^{-t})\right)=-\xi u \sin(u\xi). \quad \text{(A.4)}$$

Hence, a straightforward application of the dominated convergence theorem shows (A.3). The imaginary part of (A.2) can be treated in a similar fashion. $\square$

Lemma A.2 *For any $t \geq 0$, let Y_t be an ID random variable such that $\mathbb{E}|Y_t| < +\infty$, $\mathbb{E}Y_t = 0$, and with Lévy measure*

$$\nu_t(du)=\frac{\psi(u)-\psi(e^t u)}{|u|}du,$$

where ψ is a nonnegative function increasing on $(-\infty, 0)$ and decreasing on $(0, +\infty)$ and such that

$$\int_{|u|\leq 1}|u|\psi(u)du < \infty, \qquad \int_{|u|>1}\psi(u)du < \infty.$$

Then, for all $\xi \in \mathbb{R}$ and for all $t \in (0, 1)$,

$$\frac{1}{t}\left|\mathbb{E}e^{i\xi Y_t}-1\right|\leq C_\psi(|\xi|+|\xi|^2), \quad \text{(A.5)}$$

for some $C_\psi > 0$ only depending on ψ.

Proof Let $\xi \neq 0$ and $t \in (0, 1)$. First, since $\mathbb{E}|Y_t| < +\infty$ and since Y_t has zero mean,

$$\mathbb{E}e^{i\xi Y_t}=e^{\int_{-\infty}^{+\infty}\left(e^{iu\xi}-1-iu\xi\right)\nu_t(du)}.$$

Then,

$$\left| e^{\int_{-\infty}^{+\infty} (e^{iu\xi} - 1 - iu\xi)\nu_t(du)} - 1 \right| \leq |\xi| \max_{\omega \in [0,|\xi|]} \left| \int_{-\infty}^{+\infty} u \left(e^{iu\omega} - 1 \right) \nu_t(du) \right|.$$

Moreover, for $\omega \in [0, |\xi|]$

$$\left| \int_{-\infty}^{+\infty} u \left(e^{iu\omega} - 1 \right) \nu_t(du) \right| \leq |\omega| \int_{|u| \leq 1} |u|^2 \nu_t(du) + 2 \int_{|u| > 1} |u| \nu_t(du).$$

Let us bound the two terms $\int_{|u| \leq 1} |u|^2 \nu_t(du)$ and $\int_{|u| > 1} |u| \nu_t(du)$ present above. We only consider these integrals on $(0, +\infty)$ since similar arguments provide the same type of bounds on $(-\infty, 0)$. For the first term,

$$\begin{aligned}
\int_0^1 u \left(\psi(u) - \psi(e^t u) \right) du &= \int_0^1 u\psi(u)du - \int_0^1 u\psi(e^t u)du \\
&= \int_0^1 u\psi(u)du - e^{-2t} \int_0^{e^t} u\psi(u)du \\
&= -e^{-2t} \int_1^{e^t} u\psi(u)du + (1 - e^{-2t}) \int_0^1 u\psi(u)du \\
&\leq \sup_{1 < u < e} |u\psi(u)|(e^t - 1) + (1 - e^{-2t}) \int_0^1 u\psi(u)du.
\end{aligned}$$

Thus,

$$\int_{|u| \leq 1} |u|^2 \nu_t(du) \leq C_\psi((e^t - 1) + (1 - e^{-2t})).$$

for some constant $C_\psi > 0$ only depending on ψ. For the second term,

$$\begin{aligned}
\int_1^{+\infty} (\psi(u) - \psi(e^t u))du &= \int_1^{+\infty} \psi(u)du - e^{-t} \int_{e^t}^\infty \psi(u)du \\
&\leq \int_1^{e^t} \psi(u)du + (1 - e^{-t}) \int_1^\infty \psi(u)du \\
&\leq C_\psi \left((e^t - 1) + (1 - e^{-t}) \right).
\end{aligned}$$

The conclusion of the lemma then follows, ensuring that (A.5) holds true. $\square$

Lemma A.3 *Let $r \geq 1$, and let X and Y be two random variables. Then,*

$$d_{W_r}(X, Y) = \sup_{h \in C_c^\infty(\mathbb{R}) \cap \mathcal{H}_r} |\mathbb{E}\, h(X) - \mathbb{E}\, h(Y)|. \tag{A.6}$$

Proof Let $r \geq 1$ and let X and Y be two random variables with respective laws μ_X and μ_Y. First, note that

$$d_{W_r}(X, Y) \geq \sup_{h \in C_c^\infty(\mathbb{R}) \cap \mathcal{H}_r} |\mathbb{E}\, h(X) - \mathbb{E}\, h(Y)|.$$

Now, let $h \in \mathcal{H}_r$ and let $(h_\varepsilon)_{\varepsilon>0}$ be a regularization of h by convolution such that

$$\|h - h_\varepsilon\|_\infty \leq \varepsilon, \quad \|h_\varepsilon^{(k)}\|_\infty \leq 1, \quad 0 \leq k \leq r.$$

Let $\phi \in C_c^\infty(\mathbb{R})$ be an even function with values in $[0, 1]$ and such that $\phi(x) = 1$, for $x \in [-1, 1]$. Let $M \geq 1$ and set $\phi_M(x) = \phi(x/M)$, for all $x \in \mathbb{R}$. Next, denote by $h_{M,\varepsilon}$ the $C_c^\infty(\mathbb{R})$ function defined, for all $x \in \mathbb{R}$, by

$$h_{M,\varepsilon}(x) = \phi_M(x) h_\varepsilon(x).$$

Then,

$$
\begin{aligned}
|\mathbb{E}\, h(X) - \mathbb{E}\, h(Y)| &\leq |\mathbb{E}\, h_{M,\varepsilon}(X) - \mathbb{E}\, h_{M,\varepsilon}(Y)| + \mathbb{E}\,|h(X) - h_{M,\varepsilon}(X)| + \mathbb{E}\,|h(Y) - h_{M,\varepsilon}(Y)| \\
&\leq |\mathbb{E} h_{M,\varepsilon}(X) - \mathbb{E} h_{M,\varepsilon}(Y)| + \mathbb{E}\,|h(X) - h_\varepsilon(X)| + \mathbb{E}\,|h_\varepsilon(X) - h_{M,\varepsilon}(X)| \\
&\quad + \mathbb{E}\,|h(Y) - h_\varepsilon(Y)| + \mathbb{E}\,|h_\varepsilon(Y) - h_{M,\varepsilon}(Y)| \\
&\leq |\mathbb{E} h_{M,\varepsilon}(X) - \mathbb{E} h_{M,\varepsilon}(Y)| + 2\varepsilon + \mathbb{E}\,|h_\varepsilon(X) - h_{M,\varepsilon}(X)| + \mathbb{E}\,|h_\varepsilon(Y) - h_{M,\varepsilon}(Y)| \\
&\leq |\mathbb{E} h_{M,\varepsilon}(X) - \mathbb{E} h_{M,\varepsilon}(Y)| + 2\varepsilon + \int_{\mathbb{R}} |1 - \phi_M(x)|\,\mu_X(dx) + \int_{\mathbb{R}} |1 - \phi_M(y)|\,\mu_Y(dy).
\end{aligned}
$$

Now, choosing $M \geq 1$ large enough so that

$$\int_{\mathbb{R}} |1 - \phi_M(x)|\,\mu_X(dx) + \int_{\mathbb{R}} |1 - \phi_M(y)|\,\mu_Y(dy) \leq 2\varepsilon,$$

it follows that, for such $M \geq 1$,

$$|\mathbb{E} h(X) - \mathbb{E} h(Y)| \leq |\mathbb{E} h_{M,\varepsilon}(X) - \mathbb{E} h_{M,\varepsilon}(Y)| + 4\varepsilon.$$

Moreover, for all $x \in \mathbb{R}$, for all $M \geq 1$ and for all $\varepsilon > 0$

$$|h_{M,\varepsilon}(x)| \leq 1,$$

while, for all $x \in \mathbb{R}$ and for all $1 \leq k \leq r$,

$$
\begin{aligned}
|h_{M,\varepsilon}^{(k)}(x)| &\leq \sum_{p=0}^{k} \binom{k}{p} |h_\varepsilon^{(k-p)}(x)||\phi_M^{(p)}(x)| \\
&\leq \left(1 + C_k \sum_{p=1}^{k} \frac{1}{M^p} \right),
\end{aligned}
$$

for some $C_k > 0$ which only depends on k and on ϕ. Thus,

$$|\mathbb{E}h(X) - \mathbb{E}h(Y)| \leq \left(1 + C_r \sum_{p=1}^{r} \frac{1}{M^p}\right) \sup_{h \in C_c^\infty(\mathbb{R}) \cap \mathcal{H}_r} |\mathbb{E}h(X) - \mathbb{E}h(Y)| + 4\varepsilon,$$

for some appropriate constant $C_r > 0$ only depending on $r > 0$ and on ϕ. Letting first $M \to +\infty$ and then $\varepsilon \to 0^+$ gives the result. $\qquad\square$

Lemma A.4 *Let X and Y be two random variables. Then, for $r \geq 2$,*

$$d_{W_{r-1}}(X, Y) \leq 3\sqrt{2}\sqrt{d_{W_r}(X, Y)}.$$

Moreover,

$$d_{W_1}(X, Y) \leq \left(3\sqrt{2}\right)^{\sum_{k=1}^{r-1} \frac{1}{2^{k-1}}} \left(d_{W_r}(X, Y)\right)^{\frac{1}{2^{r-1}}},$$

for $r \geq 2$.

Proof Let $r \geq 2$, let h be an element of $\mathcal{H}_{r-1}$ and let $\varepsilon > 0$. Assume that $d_{W_r}(X, Y) \neq 0$. Let h_ε be defined, for all $x \in \mathbb{R}$, by

$$h_\varepsilon(x) := \int_{-\infty}^{+\infty} h(x - y) \exp\left(-\frac{y^2}{2\varepsilon^2}\right) \frac{dy}{\sqrt{2\pi}\varepsilon}.$$

Then, for all $0 \leq k \leq r - 1$,

$$\|h_\varepsilon^{(k)}\|_\infty \leq 1 \text{ and } \|h - h_\varepsilon\|_\infty \leq \varepsilon.$$

Moreover, by an integration by parts, $\|h_\varepsilon^{(r)}\|_\infty \leq 1/\varepsilon$. Thus,

$$|\mathbb{E}h(X) - \mathbb{E}h(Y)| \leq 2\varepsilon + |\mathbb{E}h_\varepsilon(X) - \mathbb{E}h_\varepsilon(Y)|.$$

Choosing $\varepsilon \in (0, 1)$ implies

$$|\mathbb{E}h(X) - \mathbb{E}h(Y)| \leq 2\varepsilon + \frac{1}{\varepsilon} d_{W_r}(X, Y).$$

Now, taking $\varepsilon = \sqrt{d_{W_r}(X, Y)/(2(1 + d_{W_r}(X, Y)))}$, which clearly belongs to $(0, 1)$, leads to

$$|\mathbb{E}h(X) - \mathbb{E}h(Y)| \leq \sqrt{\frac{2d_{W_r}(X, Y)}{1 + d_{W_r}(X, Y)}} + \sqrt{2\left(1 + d_{W_r}(X, Y)\right) d_{W_r}(X, Y)}.$$

Finally, since $d_{W_r}(X, Y) \leq 2$ and since $1/(1 + d_{W_r}(X, Y)) \leq 1$,

$$|\mathbb{E}h(X) - \mathbb{E}h(Y)| \le \sqrt{2d_{W_r}(X, Y)} + \sqrt{6d_{W_r}(X, Y)},$$

which implies that

$$d_{W_{r-1}}(X, Y) \le 3\sqrt{2}\sqrt{d_{W_r}(X, Y)}.$$

A recursive argument concludes the proof.   □

Lemma A.5 *Let X and Y be two random variables. Then,*

$$d_{W_1}(X, Y) = d_{FM}(X, Y).$$

Proof First, since $\mathcal{H}_1 \subset BLip(1)$,

$$d_{W_1}(X, Y) \le d_{FM}(X, Y).$$

Let h be in $BLip(1)$, let $\varepsilon > 0$ and let h_ε be a regularization of h by convolution such that

$$\|h_\varepsilon\|_\infty \le 1, \qquad \|h - h_\varepsilon\|_\infty \le \varepsilon, \qquad \|h_\varepsilon\|_{Lip(1)} \le 1.$$

Then,

$$|\mathbb{E}h(X) - \mathbb{E}h(Y)| \le 2\varepsilon + |\mathbb{E}h_\varepsilon(X) - \mathbb{E}h_\varepsilon(Y)|$$
$$\le 2\varepsilon + d_{W_1}(X, Y),$$

since h_ε belongs to $\mathcal{H}_1$. Then, letting $\varepsilon \to 0$ gives

$$d_{FM}(X, Y) \le d_{W_1}(X, Y),$$

which concludes the proof of the lemma.   □

The next lemma shows that the Kolmogorov distance and the smooth Wasserstein-1 distance are naturally related.

Lemma A.6 *Let X be a random variable with a bounded density h_X and let Y be a random variable. Then,*

$$d_K(X, Y) \le \left(1 + \frac{\|h_X\|_\infty}{2}\right)\sqrt{d_{W_1}(X, Y)}.$$

Proof If $d_{W_1}(X, Y) = 0$, there is nothing to prove. If $d_{W_1}(X, Y) \ge 1$, then since $d_K(X, Y) \le 1$,

$$d_K(X, Y) \le 1 \le \left(1 + \frac{\|h_X\|_\infty}{2}\right)\sqrt{d_{W_1}(X, Y)}.$$

So, let us assume that $0 < d_{W_1}(X, Y) < 1$. Let $x \in \mathbb{R}$ and let g_x be the indicator function of the set $(-\infty, x]$. Let $\varepsilon \in (0, 1)$ and $g_{x,\varepsilon}$ be the continuous function which is equal to 1 on $(-\infty, x]$, to 0 on $[x + \varepsilon, +\infty)$ and which is linear and decreasing in between. Then,

$$\mathbb{E}g_x(Y) - \mathbb{E}g_x(X) = \mathbb{E}g_x(Y) - \mathbb{E}g_{x,\varepsilon}(X) + \mathbb{E}g_{x,\varepsilon}(X) - \mathbb{E}g_x(X)$$
$$\leq \mathbb{E}g_{x,\varepsilon}(Y) - \mathbb{E}g_{x,\varepsilon}(X) + \mathbb{E}g_{x,\varepsilon}(X) - \mathbb{E}g_x(X).$$

Let us start with the second difference. By definition and the boundedness of the density of X,

$$\mathbb{E}g_{x,\varepsilon}(X) - \mathbb{E}g_x(X) = \int_{\mathbb{R}} \left(g_{x,\varepsilon}(y) - g_x(y)\right) h_X(y) dy \leq \|h_X\|_\infty \frac{\varepsilon}{2}.$$

Now, note

$$\|g_{x,\varepsilon}\|_\infty \leq 1, \quad \|g_{x,\varepsilon}^{(1)}\|_\infty \leq \varepsilon^{-1},$$

so that the function $\varepsilon g_{x,\varepsilon}$ belongs to $BLip(1)$ since $\varepsilon \in (0, 1)$. Therefore, by Lemma A.5,

$$\mathbb{E}g_{x,\varepsilon}(Y) - \mathbb{E}g_{x,\varepsilon}(X) \leq \frac{1}{\varepsilon} d_{W_1}(Y, X).$$

Thus,

$$\mathbb{E}g_x(Y) - \mathbb{E}g_x(X) \leq \frac{1}{\varepsilon} d_{W_1}(X, Y) + \|h_X\|_\infty \frac{\varepsilon}{2}.$$

Now, taking $\varepsilon = \sqrt{d_{W_1}(X, Y)} < 1$ gives,

$$\mathbb{E}g_x(Y) - \mathbb{E}g_x(X) \leq \left(1 + \frac{\|h_X\|_\infty}{2}\right) \sqrt{d_{W_1}(X, Y)}.$$

The same inequality holds for $\mathbb{E}g_x(X) - \mathbb{E}g_x(Y)$, using the continuous function $\tilde{g}_{x,\varepsilon}$ which is equal to 1 on $(-\infty, x - \varepsilon]$, to 0 on $[x, +\infty)$ and which is linear and decreasing in between. $\qquad\square$

Lemma A.7 *Let* $(p_n)_{n \geq 1}$ *be an increasing enumeration of the prime numbers starting at 2 and let* $(X_n)_{n \geq 1}$ *be a sequence of independent random variables Bernoulli distributed such that, for each* $k \geq 1$, $\mathbb{P}(X_k = 1) = 1/(1 + p_k)$ *and* $\mathbb{P}(X_k = 0) = p_k/(1 + p_k)$. *Let* $(S_n)_{n \geq 1}$ *be given by,* $S_n = \sum_{k=1}^n (\ln p_k) X_k / \ln p_n$, *for all* $n \geq 1$, *and let* I *be a discrete random variable independent of* $(X_n)_{n \geq 1}$ *and with values in* $\{1, ..., n\}$, *such that, for all* $k \in \{1, ..., n\}$,

$$\mathbb{P}(I = k) = \frac{\ln p_k}{\ln p_n \mathbb{E}S_n(1 + p_k)}.$$

Finally, let U be a uniform random variable on [0, 1] independent of $(X_n)_{n \geq 1}$. Then, for all $n \geq 2$

$$\left| 1 - \frac{1}{\ln p_n} \sum_{k=1}^{n} \frac{\ln p_k}{1 + p_k} \right| \leq \frac{C_1}{\ln n}$$

for some $C_1 > 0$ independent of n. Moreover, there exists a well-chosen coupling between U and I such that, for all $n \geq 2$

$$\mathbb{E} \left| U - \frac{\ln p_I}{\ln p_n} \right| \leq \frac{C_2}{\ln n}, \tag{A.7}$$

for some $C_2 > 0$ independent of n.

Proof We divide the proof into two steps.
Step 1: To start, let us bound the term $|\mathbb{E}S_n - 1|$, for all $n \geq 2$. First,

$$\mathbb{E}S_n - 1 = \frac{1}{\ln p_n} \left(\sum_{k=1}^{n} \frac{\ln p_k}{(p_k + 1)} - \ln p_n \right).$$

Next by the prime number theorem [88, Sect. 1.10],

$$\ln p_n = \ln n + \mathcal{O}(|\ln \ln n|),$$

with the usual definition of $\mathcal{O}$. Moreover, by Mertens first theorem (see, e.g., [88, Proposition 1.51] or [89, Theorem 1.8])

$$\sum_{k=1}^{n} \frac{\ln p_k}{p_k + 1} - \ln p_n = \sum_{k=1}^{n} \frac{\ln p_k}{p_k} - \ln p_n - \sum_{k=1}^{n} \frac{\ln p_k}{p_k(p_k + 1)}$$
$$= \mathcal{O}(1).$$

Thus,

$$\mathbb{E}S_n - 1 = \mathcal{O}\left(\frac{1}{\ln n} \right),$$

and so we are done with Step 1.
Step 2: To continue the proof of the lemma, let us deal with (A.7). For this purpose, set $F_0 = 0$ and for $1 \leq j \leq n$,

$$F_j = \frac{1}{\mathbb{E}S_n \ln p_n} \sum_{k=1}^{j} \frac{\ln p_k}{1 + p_k}.$$

Note that $0 \le F_j \le 1$, for all $0 \le j \le n$. Now, consider the coupling between U and I defined, for all $1 \le j \le n$ by $I = j$ if $U \in [F_{j-1}, F_j]$. Note that U is independent of S_n since I is. Moreover,

$$\mathbb{E}\left|U - \frac{\ln p_I}{\ln p_n}\right| = \sum_{j=1}^{n} \mathbb{P}(I = j)\mathbb{E}\left(\left|U - \frac{\ln p_I}{\ln p_n}\right| \,\middle|\, I = j\right).$$

To continue, we need to control the following quantity:

$$q_j := \max\left(\left|F_{j-1} - \frac{\ln p_j}{\ln p_n}\right|, \left|F_j - \frac{\ln p_j}{\ln p_n}\right|\right),$$

for $1 \le j \le n$. Let us bound $\left|F_j - \frac{\ln p_j}{\ln p_n}\right|$, for $1 \le j \le n$. For all $n \ge 2$ and for all $1 \le j \le n$

$$F_j - \frac{\ln p_j}{\ln p_n} = \frac{1}{\ln p_n}\left(\sum_{k=1}^{j} \frac{\ln p_k}{1 + p_k} - \ln p_j\right) + \frac{1}{\ln p_n}\left(\frac{1}{\mathbb{E}S_n} - 1\right)\sum_{k=1}^{j} \frac{\ln p_k}{p_k + 1}.$$

Thus, using again Mertens first theorem and standard inequalities gives

$$\left|F_j - \frac{\ln p_j}{\ln p_n}\right| \le \frac{C_1}{\ln p_n} + \frac{C_2}{\ln p_n}\left|\frac{\mathbb{E}S_n - 1}{\mathbb{E}S_n}\right| + C_3\left|\frac{\mathbb{E}S_n - 1}{\mathbb{E}S_n}\right|,$$

for some constants $C_1 > 0$, $C_2 > 0$, and $C_3 > 0$ independent of n and j. Similarly, using the fact that $F_j - F_{j-1} = \mathbb{P}(I = j)$,

$$\left|F_{j-1} - \frac{\ln p_j}{\ln p_n}\right| \le \frac{C_1}{\ln p_n} + \frac{C_2}{\ln p_n}\left|\frac{\mathbb{E}S_n - 1}{\mathbb{E}S_n}\right| + C_3\left|\frac{\mathbb{E}S_n - 1}{\mathbb{E}S_n}\right| + \mathbb{P}(I = j).$$

Then, for all $1 \le j \le n$

$$q_j \le \frac{C_1}{\ln p_n} + \frac{C_2}{\ln p_n}\left|\frac{\mathbb{E}S_n - 1}{\mathbb{E}S_n}\right| + C_3\left|\frac{\mathbb{E}S_n - 1}{\mathbb{E}S_n}\right| + \mathbb{P}(I = j).$$

The previous bounds imply that

$$\mathbb{E}\left|U - \frac{\ln p_I}{\ln p_n}\right| \le C\left(\frac{1}{\ln p_n} + \frac{1}{\ln p_n}\left|\frac{\mathbb{E}S_n - 1}{\mathbb{E}S_n}\right| + \left|\frac{\mathbb{E}S_n - 1}{\mathbb{E}S_n}\right|\right) + \sum_{j=1}^{n} \mathbb{P}(I = j)^2.$$

But

$$\sum_{j=1}^{n} \mathbb{P}(I = j)^2 \leq \frac{C}{(\ln p_n)^2},$$

for some $C > 0$ independent of n.

Combining both steps together and using the prime number theorem lead to

$$\mathbb{E}\left| U - \frac{\ln p_I}{\ln p_n} \right| = \mathcal{O}\left(\frac{1}{\ln n}\right),$$

which concludes the proof of the lemma.   □

References

1. S. Albeverio, S. Rüdinger, J.L. Wu, Invariant measures and symmetry property of Lévy type operators. Potential Anal. **13**(2), 147–168 (2000)
2. D. Applebaum, *Lévy Processes and Stochastic Calculus* (Cambridge University Press, Cambridge, 2009)
3. B. Arras, C. Houdré, On Stein's method for multivariate self-decomposable laws with finite first moment. Electron. J. Probab. (2019), arXiv:1809.02050
4. B. Arras, G. Mijoule, G. Poly, Y. Swan, A new approach to the Stein-Tikhomirov method: with applications to the second Wiener chaos and Dickman convergence (2017), arXiv:1605.06819
5. R. Arratia, A.D. Barbour, S. Tavaré, *Logarithmic Combinatorial Structures: A Probabilistic Approach* (European Mathematical Society, 2003)
6. R. Arratia, P. Baxendale, Bounded size bias coupling: a Gamma function bound, and universal Dickman-function behavior. Probab. Theory Relat. Fields **162**, 411–429 (2015)
7. R. Arratia, L. Goldstein, F. Kochman, Size bias for one and all. Probab. Surv. **16**, 1–61 (2019)
8. A.D. Barbour, Stein's method and Poisson process convergence. J. Appl. Probab. **25A**, 175–184 (1988)
9. A.D. Barbour, Stein's method for diffusion approximations. Probab. Theory Relat. Fields. **84**(3), 297–322 (1990)
10. A.D. Barbour, L.H.Y. Chen, *An Introduction to Stein's Method*. Lecture Notes Series, Institute for Mathematical Sciences, National University of Singapore, vol. 4 (Singapore University Press, Singapore, 2005)
11. A.D. Barbour, L.H.Y. Chen, W.-L. Loh, Compound Poisson approximations for non-negative random variables via Stein's method. Ann. Probab. **20**(4), 1843–1866 (1992)
12. A.D. Barbour, H.L. Gan, A. Xia, Stein factors for negative binomial approximation in Wasserstein distance. Bernoulli **21**(2), 1002–1013 (2015)
13. A.D. Barbour, L. Holst, S. Janson, *Poisson Approximation* (Oxford Science Publications, Oxford)
14. O.E. Barndorff-Nielsen, T. Mikosch, S.I. Resnick, *Lévy Processes: Theory and Applications* (Springer Science and Business Media, New York, 2001)
15. J. Bartroff, L. Goldstein, Ü. Islak, Bounded size biased couplings, log concave distributions and concentration of measure for occupancy models. Bernoulli **24**(4B), 3283–3317 (2018)
16. J. Bertoin, *Lévy Processes*, vol. 121 (Cambridge University Press, Cambridge, 1998)
17. C. Bhattacharjee, L. Goldstein, Dickman approximation in simulation, summations and perpetuities. Bernoulli (2018), arXiv:16.08192v5
18. P. Billingsley, The probability theory of additive arithmetic functions. Ann. Probab. **2**(5), 749–791 (1974)

19. L. Bondesson, *Generalized Gamma Convolutions and Related Classes of Distributions and Densities* (Lectures Notes in Statistics (Springer, Berlin, 1992)
20. P. Bosch, T. Simon, A proof of Bondesson's conjecture on stable densities. Ark. Mat. **54**(1), 31–38 (2016)
21. T.C. Brown, M.J. Phillips, Negative binomial approximations with Stein's method. Methodol. Comput. Appl. Probab. **1**(4), 407–421 (1999)
22. T.C. Brown, A. Xia, Stein's method and birth-death processes. Ann. Probab. **29**, 1373–1403 (2001)
23. F. Cellarosi, Y.G. Sinai, *Non-standard limit theorems in number theory, Prokhorov and Contemporary Probability Theory* (Springer, Berlin, 2013), pp. 197–213
24. D. Chafaï, A. Joulin, Intertwining and commutation relations for birth-death processes. Bernoulli **19**(5), 1855–1879 (2013)
25. S. Chatterjee, A new method of normal approximation. Ann. Probab. **36**(4), 1584–1610 (2008)
26. S. Chatterjee, Fluctuations of eigenvalues and second order Poincaré inequalities. Probab. Theory Relat. Fields **143**, 1–40 (2009)
27. S. Chatterjee, A short survey of Stein's method. Proc. ICM IV **1–24**, (2014)
28. S. Chatterjee, P. Diaconis, E. Meckes, Exchangeable pairs and Poisson approximation. Probab. Surv. **2**, 64–106 (2005)
29. S. Chatterjee, J. Fulman, A. Röllin, Exponential approximation by Stein's method and spectral graph theory. Lat. Am. J. Probab. Math. Stat. **8**, 197–223 (2011)
30. L.H.Y. Chen, Poisson approximation for dependent trials. Ann. Probab. **3**(3), 534–545 (1975)
31. L.H.Y. Chen, L. Goldstein, Q.M. Shao, *Normal Approximation by Stein's Method Probability and Its Application* (Springer, Heidelberg, 2011)
32. Z.-Q. Chen, X. Zhang, Heat kernels and analyticity of non-symmetric jump diffusion semigroups. Probab. Theory Relat. Fields **165**, 267–312 (2016)
33. G. Christoph, W. Wolf, *Convergence Theorems with a Stable Limit Law* (Akademie-Verlag, Berlin, 1993)
34. R. Cont, P. Tankov, *Financial Modelling with Jump Processes* (Chapman and Hall/CRC, 2004)
35. P. Diaconis, S. Holmes, Stein's Method: Expository Lectures and Applications. IMS Lecture Notes-Monograph Series **46**, (2004)
36. A. Diédhiou, On the self-decomposability of the half-Cauchy distribution. J. Math. Anal. Appl. **220**, 42–64 (1998)
37. C. Döbler, G. Peccati, The Gamma Stein equation and noncentral de Jong theorems. Bernoulli **24**(4B), 3384–3421 (2018)
38. J.A. Domínguez-Molina, The Tracy-Widom distribution is not infinitely divisible. Stat. Probab. Lett. **123**, 56–60 (2017)
39. R.M. Dudley, *Real Analysis and Probability*, 2nd edn. (Cambridge University Press, Cambridge, 2002)
40. W. Ehm, Binomial approximation to the Poisson binomial distribution. Stat. Probab. Lett. **11**, 7–16 (1991)
41. S.N. Ethier, T.G. Kurtz, *Markov Processes: Characterization and Convergence*, vol. 282 (Wiley, New York, 2009)
42. J. Fulman, N. Ross, Exponential approximation and Stein's method of exchangeable pairs. Lat. Am. J. Probab. Math. Stat. **10**(1), 1–13 (2013)
43. R.E. Gaunt, A.M. Pickett, G. Reinert, Chi-square approximation by Stein's method with application to Pearson's statistic. Ann. Appl. Probab. **27**(2), 720–756 (2017)
44. S. Ghosh, L. Goldstein, Concentration of measures via size-biased couplings. Probab. Theory Relat. Fields **149**, 271–278 (2011)
45. B.V. Gnedenko, A.N. Kolmogorov, *Limit Distributions for Sums of Independent Random Variables* (Addison-Wesley Publishing Company, Cambridge, 1954)
46. L. Goldstein, Ü. Islak, Concentration inequalities via zero bias couplings. Stat. Probab. Lett. **86**, 17–23 (2014)
47. L. Goldstein, G. Reinert, Stein's method and the zero bias transformation with application to simple random sampling. Ann. Appl. Probab. **7**(4), 935–952 (1997)

48. F. Götze, On the rate of convergence in the multivariate CLT. Ann. Probab. **19**(2), 724–739 (1991)
49. C. Halgreen, Self-decomposability of the generalized inverse Gaussian and hyperbolic distributions. Z. Wahrscheinlichkeitstheorie verw. Gebiete **47**, 13–17 (1979)
50. P. Hall, Two-sided bounds on the rate of convergence to a stable law. Probab. Theory Relat. Fields **57**(3), 349–364 (1981)
51. W.J. Hall, J.A. Wellner, The rate of convergence in law of the maximum of an exponential sample. Stat. Neerlandica **33**, 151–154 (1979)
52. K.H. Hofmann, Z.J. Jurek, Some analytic semigroups occurring in probability theory. J. Theor. Probab. **9**(3), 745–763 (1996)
53. C. Houdré, Remarks on deviation inequalities for functions of infinitely divisible random vectors. Ann. Probab. **30**(3), 1223–1237 (2002)
54. C. Houdré, P. Marchal, On the concentration of measure phenomenon for stable and related random vectors. Ann. Probab. **32**(2), 1496–1508 (2004)
55. C. Houdré, V. Pérez-Abreu, D. Surgailis, Interpolation, correlation identities and inequalities for infinitely divisible variables. J. Fourier Anal. Appl. **4**(6), 651–668 (1998)
56. S. Janson, *Gaussian Hilbert Spaces*, vol. 129 (Cambridge University Press, Cambridge, 1997)
57. A. Janssen, D.M. Mason, On the rate of convergence of sums of extremes to a stable law. Probab. Theory Relat. Fields **86**(2), 253–264 (1990)
58. Z.J. Jurek, On relations between Urbanik and Mehler semigroups. Probab. Math. Stat. **29**(2), 297–308 (2009)
59. A.Ya. Khintchine, *Limit Laws for Sums of Independent Random Variables* (ONTI, Moscow-Leningrad (in Russian, 1938)
60. R. Kuske, J.B. Keller, Rate of convergence to a stable law. SIAM J. Appl. Math. **61**(4), 1308–1323 (2000)
61. M. Ledoux, I. Nourdin, G. Peccati, Stein's method, logarithmic Sobolev inequality and transport inequalities. Geom. Funct. Anal. **25**, 256–306 (2015)
62. P. Lévy, *Théorie de l'addition des variables aléatoires*, 1st edn. (Gauthier-Villars, Paris, 1937)
63. C. Ley, G. Reinert, Y. Swan, Stein's method for comparison of univariate distributions. Probab. Surv. **14**, 1–52 (2017)
64. G.D. Lin, C.-Y. Hu, The Riemann zeta distribution. Bernoulli **7**(5), 817–828 (2001)
65. H.M. Luk, Stein's method for the Gamma distribution and related statistical applications, University of Southern California, 1994
66. E. Lukacs, *Characteristic Functions*, 2nd edn. (Griffin, London, 1970), p. 1820
67. M.B. Marcus, J. Rosinski, L^1-norm of infinitely divisible random vectors and certain stochastic integrals. Electron. Commun. Probab. **6**, 15–29 (2001)
68. I.W. McKeague, E. Peköz, Y. Swan, Stein's method and approximating the quantum harmonic oscillator. Bernoulli (to appear)
69. T. Nakamura, A modified Riemann zeta distribution in the critical strip. Proc. Am. Math. Soc. **143**(2), 897–905 (2015)
70. T. Nakamura, A complete Riemann zeta distribution and the Riemann hypothesis. Bernoulli **21**(1), 604–617 (2015)
71. I. Nourdin, G. Peccati, Stein's method on Wiener chaos. Probab. Theory Relat. Fields **145**, 75–118 (2009)
72. I. Nourdin, G. Peccati, *Normal Approximations with Malliavin Calculus: From Stein's Method to Universality* (Cambridge University Press, Cambridge, 2012)
73. I. Nourdin, G. Poly, Convergence in law in the second Wiener/Wigner chaos. Electron. Commun. Probab. **36**, 1–12 (2012)
74. F.W.J. Olver, D.W. Lozier, R.F. Boisvert, C.W. Clark, *NIST Handbook of Mathematical Functions* (Cambridge University Press, Cambridge, 2010)
75. J.R. Partington, *Linear Operators and Linear Systems: An Analytical Approach to Control Theory* (Cambridge University Press, Cambridge, 2004)
76. E.A. Peköz, A. Röllin, New rates for exponential approximation and the theorems of Rényi and Yaglom. Ann. Probab. **39**(2), 587–608 (2011)

77. V.V. Petrov, *Limit Theorems of Probability Theory* (Oxford University Press, Oxford, 1995)
78. A. Pickett, Rates of convergence of Chi-square approximations via Steins method, Ph.D. thesis, University of Oxford, 2004
79. J. Pike, H. Ren, Stein's method and the Laplace distribution. ALEA Lat. Am. J. Probab. Math. Stat. **11**(1), 571–587 (2014)
80. D. Revuz, M. Yor, *Continuous Martingales and Brownian Motion*, vol. 293, 3rd edn. (Springer, Berlin, 1999)
81. N.F. Ross, Fundamentals of Stein's method. Probab. Surv. **8**, 210–293 (2011)
82. N.F. Ross, Power laws in preferential attachment graphs and Stein's method for the negative binomial distribution. Adv. Appl. Prob. **45**, 876–893 (2013)
83. S.G. Samko, A.A. Kilbas, O.I. Marichev, *Fractional Integrals and Derivatives* (Gordon and Breach Science Publishers, Yverdon, 1993)
84. K-I. Sato, *Lévy Processes and Infinitely Divisible Distributions* (Cambridge University Press, Corrected Printing with Supplements, 2015)
85. C. Stein, A bound for the error in the normal approximation to the distribution of a sum of dependent random variables, in *Proceedings of the Sixth Berkeley Symposium on Mathematical Statistics and Probability* (1972), pp. 583–602
86. C. Stein, Approximate Computation of Expectations. Institute of Mathematical Statistics Lecture Notes Monograph Series **7**, (1986)
87. F.W. Steutel, K. Van Harn, *Infinite Divisibility of Probability Distributions on the Real Line* (CRC Press, 2003)
88. T. Tao, V.H. Vu, *Additive Combinatorics*, vol. 105 (Cambridge University Press, Cambridge, 2006)
89. G. Tenenbaum, *Introduction to Analytic and Probabilistic Number Theory*. Graduate Studies in Mathematics, vol. 163, 3rd edn. (2015)
90. C.A. Tracy, H. Widom, Distribution functions for largest eigenvalues and their applications. Proc. ICM I **587–596**, (2002)
91. A.N. Tikhomirov, On the convergence rate in the central limit theorem for weakly dependent random variables. Theory Probab. Math. Stat. **25**(4), 790–809 (1981)
92. M. Veillette, M.S. Taqqu, Properties and numerical evaluation of the Rosenblatt distribution. Bernoulli **19**(3), 982–1005 (2013)
93. C. Villani, *Optimal Transport, Old and New*, vol. 338 (Springer, Berlin, 2009)
94. L. Xu, Approximation of stable law in Wasserstein-1 distance by Stein's method. Ann. Appl. Probab. (2019), arXiv:1709.00805v3

Index

© The Author(s), under exclusive license to Springer Nature Switzerland AG 2019
B. Arras and C. Houdré, *On Stein's Method for Infinitely Divisible Laws with Finite First Moment*, SpringerBriefs in Probability and Mathematical Statistics,
https://doi.org/10.1007/978-3-030-15017-4

Printed in the United States
By Bookmasters